Friedrich Semeleder, Edward T Caswell

Rhinoscopy and Laryngoscopy

Their Value in Practical Medicine

Friedrich Semeleder, Edward T Caswell

Rhinoscopy and Laryngoscopy
Their Value in Practical Medicine

ISBN/EAN: 9783337371739

Printed in Europe, USA, Canada, Australia, Japan

Cover: Foto ©berggeist007 / pixelio.de

More available books at **www.hansebooks.com**

RHINOSCOPY

AND

LARYNGOSCOPY;

THEIR VALUE IN PRACTICAL MEDICINE.

BY

DR. FRIEDERICH SEMELEDER,

Physician in ordinary to His Majesty the Emperor of Mexico, Member of the Royal Medical
Society of Vienna, and of the Medical Society of the Pantheon in Paris. Formerly Member of the Medical Faculty of the University of
Vienna, and Surgeon to the Branch Hospital
at Gumpendorf.

Translated from the German
By EDWARD T. CASWELL, M.D.

With Woodcuts and Two Chromo-Lithographic Plates.

NEW YORK:
WILLIAM WOOD & CO., 61 WALKER STREET.
1866.

TRANSLATOR'S PREFACE.

THE accompanying work is presented to the American public with the hope that it will stimulate the medical men of our country to become acquainted with the instruments which it describes, and to reap in their daily practice the benefits of the knowledge which it imparts. Our medical literature offers but very little on these subjects, and the translator is sure that every practitioner will find this volume a valuable manual of reference and instruction in the diseases of the throat and of the nose. It comprises two separate monographs published by Dr. Semeleder, the one in March, and the other in December, 1862. The translator thought it would be well to combine them, and in doing so he has omitted some repetitions, which necessarily occurred in two treatises upon subjects so closely related; others still remain. He has retained the headings to the sections, however, for the sake of convenient reference. In some cases a want of uniformity in the author's mode of expression has been rectified. A few brief Notes and an Appendix have been added, and may not prove uninteresting to the reader. The Bibliographic List has been enlarged, and made to include, as far as possible, all the works relating to these subjects hitherto published. He may be pardoned for adding that the task has been performed with additional pleasure and zeal from the recollection of the many happy hours passed with his good friend, and faithful instructor, the author.

Providence, October 1st, 1865.

TABLE OF CONTENTS.

PART I.–RHINOSCOPY.

CHAPTER I.

METHOD OF PRACTISING RHINOSCOPY.

CHAPTER II.

PRACTICAL APPLICATION OF RHINOSCOPY.

PART II.–LARYNGOSCOPY.

CHAPTER I.

METHOD OF PRACTISING LARYNGOSCOPY.

AUTHOR'S PREFACE.

I DESPISE hypocritical words, and will offer no excuse for venturing to present the accompanying work to the judgment of the medical world. I thank my colleagues at home and abroad most heartily for their friendly support, and for their contributions, and beg that they will continue them to me in the future. They must judge whether I have made a good use of them. The volume has not, I hope, been rendered less valuable by that which I have contributed from my own experience. The work was intended to present facts and to be of practical use; how far it has fulfilled this intention the judgment of adepts in this department must decide.

The author and the publisher will soon know whether they were in error in thinking that they were presenting something demanded by the times. To have given a few illustrations would have been of but little value; many would have disproportionately increased the price of the work.

SEMELEDER.

Vienna, December, 1862

BIBLIOGRAPHIC LIST

OF

WORKS RELATING TO RHINOSCOPY AND LARYNGOSCOPY.

————o————

1. CZERMAK.—" Ueber den Kehlkopfspiegel." (The Laryngo-
scope.) Wiener med. Wochenschrift, No. 13 and 16, 1858.—
" Ueber die Inspection des Cav. pharyngo-nasale," etc. (The
Inspection of the nasal cavity, etc.). The same, No. 32, 1859.—
Second article. The same. No. 17, 1860.—" Der Kehlkopfspiegel."
(The Laryngoscope.) Leipzig, Engelmann, 1860. Second edition
revised and enlarged, 1863. (Tr.)—" Zur Verwerthung des Liston-
Garcia'schen Prinzips." (The value of Liston's and Garcia's principle.) Wiener mediz. Wochenschrift, Nos. 6 and 7, 1861.—"Appli-
cation de la photographie à la laryngoscopie et à la rhinoscopie."
Comptes rendus, Nov. 25, 1861.—" Laryngoscopische und Rhino-
scopische Mittheilungen." (Communications upon Laryngoscopy
and Rhinoscopy.) Virchow's Archiv, Bd. xxiii. 1862.

2. SEMELEDER.—" Ueber die Untersuchung des Nasenrachen-
raumes." (The examination of the naso-pharyngeal space.) Zeitschr. d. Gesellsch. d. Aerzte zu Wien, 1860, No. 19, and " Zur Rhinoscopie " (Upon Rhinoscopy), No. 47.—" Der Katheterismus der
Eustach'schen Ohrtrompete und das Rhinoscop." (The Catheterization of the Eustachian tube, and the Rhinoscope.) Oesterr.
Zeitschr. f. prakt. Heilkunde, No. 21 and 22, 1860 ; still further a
case reported in No. 27, 1860, der Wiener Allgem. med. Zeitung.

3. STÖRK.—" Rhinoscopie," Zeitschr. d. Gesellsch. d. Aerzte zu
Wien, No. 26, 1860.

4. TURCK.—" Beiträge zur Laryngoscopie und Rhinoscopie."
(Contributions to Laryngoscopy and Rhinoscopy.) Zeitschr. der
Gesellsch. d. Aerzte zu Wien, No. 21, 1860.—" Notizen zur Rhi-
noscopie." (Notes upon Rhinoscopy.) Wiener Allg. med. Zeitung,

No. 33, 1860. Moreover, in the same paper, 1861, Nos. 28, 32, 46, and 48, "Ueber Syphilitische Geschwüre im Nasenrachenraum." (Upon Syphilitic ulcerations in the naso-pharyngeal cavity); finally, "Praktische Anleitung zur Laryngoscopie." (Practical Introduction to Laryngoscopy.) Wien, 1860.—"Clinical Researches on different diseases of the Larynx, etc., examined by the Laryngoscope. Paris, Baillière, 1862. (Tr.)

5. VOLTOLINI.—"Die Besichtigung der Tuba Eustachii und der übrigen Theile des cavuum-pharyngo-nasale mittelst des Schlundkopfspiegels." (The examination of the Eustachian tube and of the other parts of the naso-pharyngeal cavity by means of the pharyngeal mirror.) Deutsche Klinik, No. 21, 1860.—In the same, "Rhinosk. Befund bei einem Schwerhörigen." (Rhinoscopic condition in a deaf person.) No. 42, 1861.—"Die Pharyngoscopie und ihre Verwerthung für die Ohrenheilkunde." (Pharyngoscopy and its worth in Aural Therapeutics.) Virchow's Archiv, Bd. xxi, Heft i.—"Der Katheterismus der Tuba Eustachii, und der Pharynxspiegel." (The catheterization of the Eustachian tube and the pharyngeal mirror.) Jahrb. der Gesellschaft der Aerzte zu Wien, ii, p. 93, 1861.—Finally, a monograph upon the occasion of the semi-centenarian Jubilee of the University of Breslau, Aug. 3, 1861.

6. DAUSCHER.—Zeitschrift der Gesellschaft der Aerzte zu Wien, No. 38, 1860.

7. ZSIGMONDY.—"Neue Folge galvanokaust. Operationen." (New results of galvano-caustic operations.) Oesterr. Zeitsch. f. prakt. Heilkunde, No. 39, 1860.

8. GERHARDT.—"Zur Anwendung des Kehlkopfspiegels." (The Use of the Laryngoscope.) Würzbürger med. Zeitschrift, 1860, iii.

9. GERHARDT v. ROTH.—"Ueber syphil. Krankheiten des Kehlkopfs." (Syphilitic diseases of the Larynx.) Virchow's Archiv, Bd. xxi.

10. MOURA-BOUROUILLOU.—"Cours complet de Laryngoscopie." Delahaye, Paris, 1861.

11. MERKEL.—"Die Functionen des Schlund- und Kehlkopfes." (The functions of the Pharynx and Larynx.) Leipzig, O. Wigand, 1862.

12. WAGNER, of New York.—"Zur Laryngoscopie und Rhino-scopie." (Laryngoscopy and Rhinoscopy.) Oesterr. Zeitsch. f. prakt. Heilkunde, No. 6, 1862.

The Translator takes the liberty of adding

13. LEWIN.—"Die Laryngoscopie. Beiträge zu ihrer Verwer-thung für praktische Medicin." (Laryngoscopy. A contribution to its value in practical medicine.) Berlin, Hirschwald, 1860.— "Klinik der Krankheiten des Kehlkopfes." i. Band," etc. (Clinic of the Diseases of the Larynx, Vol. i.) Also under the title "Inhalation Treatment of Diseases of the Respiratory Organs, with special reference to those Diseases of the Larynx made known by the Laryngoscope." With 25 wood cuts. Second enlarged and revised edition. 8vo. 563 pp. Berlin, 1865.

14. BATTAILLE. — "Nouvelles Recherches sur la phonation." Paris, Victor Masson, 1861.

15. FAUVEL.—"Du Laryngoscope au point de vue pratique." Paris, V. Masson, 1861.

16. JAMES.—"Sore Throat; its Nature, Varieties, and Treat-ment; including the use of the Laryngoscope as an aid to Diagno-sis." London, 1861.

17. YEARSLEY.—"Introduction to the Art of Laryngoscopy." London, Churchill, 1862.

18. BRUNS, V.—"Die erste Ausrottung eines Polypen in der Kehlköpfshöhle durch Zerschneiden." (The first extirpation of a polyp in the cavity of the Larynx, by cutting.) Tübingen, Laupp, 1862. Supplement to the same. do. 1863.—"Die Laryngoscopie und Laryngoscopische Chirurgie." (The Laryngoscope and Laryn-goscopic Surgery.) Lex. 8, pp. 451. Tübingen, 1865. Atlas to the same. Fol. maps, 8 plates. (Two colored.)

19. TOBOLD.—"Lehrbuch der Laryngoscopie und des local-ther-apeutischen Verfahrens bei Kehlkopf-Krankheiten." (Manual of Laryngoscopy, and of local therapeutics in diseases of the Larynx.) Berlin, Hirschwald, 1863.

20. GIBB.—"The Laryngoscope. Illustrations of its practical application, and description of its mechanism." London, Churchill, 1863.—" On Diseases of the Throat and Windpipe as reflected by the Laryngoscope ; a complete manual upon their Diagnosis and Treatment." London, Churchill, 1864.

21. JOHNSON.—Two Lectures on the Laryngoscope. The Lancet, August, 1864. (Published also in Pamph.)

22. ELSBERG.—"Laryngoscopal Medication ; or the Local Treatment of the Diseases of the Throat, Larynx, and neighboring organs under sight." New York, Wood, 1864.

23. GUILLAUME.—"Essai sur la laryngoscopie et la rhinoscopie." Paris, Delahaye, 1864.

24. MOURA-BOUROUILLOU.—"Traité pratique de laryngoscopie et de rhinoscopie, suivi d'observations." 8vo. avec pl. A. Delahaye, Paris, 1865.

25. BAUMGÄRTNER, Dr. J.—"Die Krankheiten des Kehlkopfes und deren Behandlung," etc. (The Diseases of the Larynx and their Treatment, together with a notice of some new Inhalation Apparatus and an Introduction to Laryngoscopic examination.) 8vo. 130 pp., with wood cuts. Freiburg, 1865.

26. MACKENZIE MORRELL.—"The Use of the Laryngoscope in Diseases of the Throat, with an Appendix on Rhinoscopy." 8vo. pp. 152. London, 1865.

27. DIXON, THOMAS.—"On Diseases of the Throat ; their new Treatment by the aid of the Laryngoscope." Post 8vo. pp. 102. London, 1865.

PART I.
RHINOSCOPY.

CHAPTER I.

Its History and Literature.

ALTHOUGH Bozzini at the commencement of the present century, in a work long since forgotten, Der Lichtleiter, etc., Weimar, 1807, proposed to examine the parts "behind the hanging palate"—although a passage in a work of the celebrated aurist R. Wilde, of Dublin, shows that he imagined that we might examine the openings of the Eustachian tube from the bucco-pharyngeal cavity by the aid of a small mirror; yet the actual history and development of a method of examining the nasal portion of the pharynx and the nasal cavities, by the use of small mirrors, begin with Professor J. Czermak in 1858. He first saw and demonstrated upon the living, the openings of the posterior nares, and the pharyngeal openings of the ear-trumpet; established by ocular demonstration the feasibility of this procedure so violently attacked "on anatomical grounds;" gave to the same the name of rhinoscopy; and has won for it laborers, and created for it a future.

For about two years the cultivation of the laryngoscopic methods engrossed the attention of few physicians, save that small number who become particularly enthusiastic upon such novelties. In the spring of 1860, circumstances contributed to give a new impulse to rhinoscopy. Soon after Czermak had given an account of the rhinoscopic method, in the Vienna *Mediz. Wochenschrift*, 1859, No. 32, and in his work upon the Laryngoscope and its value in Physiology and Medicine (Der Kehlkopfspiegel, und seine Verwerthung für Physiologie und Medizin), Leipzig, Engelmann, 1860, he published (in the Vienna *Mediz. Wochenschrift*, 1860, No. 17) the first report of a patholo-

gical case in the naso-pharyngeal space, and then a whole cata
logue of publications quickly followed.*

These publications include the rhinoscopic method in general
and in particular; some of them, and those the more recent,
report cases of the practical value of rhinoscopy in order to
stimulate . the practice of this method of examination. By
chance, and by the great number of his examinations, the author
was fortunately enabled to report many peculiar and many
pathological cases; he also kindly requested his colleagues to furnish him with any pertinent observations. The slight success of
such an invitation, as well as a glance at the medical journals,
show that rhinoscopy, under which term we include the examination of all parts of the naso-pharyngeal cavity, has really
found but a very limited number of investigators, and that it is
still, in general, greatly undervalued. Although Czermak, by
demonstrations upon himself in London and Paris, gained something for rhinoscopy, and although his works since that time
have been translated into French and English;† although some
of the author's articles were translated and reprinted in the
French and English journals; and though he had in a course at
La Charité in Paris, instructed some young physicians in the
mode of proceeding; yet, according to the observation and
report of Dr. Fürstenberg, after a journey through Germany,
France, and England, Rhinoscopy had gained no foothold outside of Germany; and yet the results accomplished show that
this method has its own indisputable, though limited value in
medical practice. In particular, aural surgery has much to
expect from it. There is no doubt but that the difficulty of the
procedure contributes materially to this neglect. Although
Voltolini now practises it so easily, he forgets how great patience
he had, and how many useless attempts he made, in order to
arrive at his facility. The many practical courses which the
author has given, show him clearly that the rhinoscopic method
is far more difficult than the laryngoscopic; and such, indeed,
is naturally the case.

---

* A list of all the publications bearing upon Rhinoscopy and known to the author
was here given in a foot-note. For convenience' sake it has been placed at the
commencement of this volume. (Tr.)

† Czermak: Du Laryngoscope. J. B. Baillière, Paris, 1860. On the Laryngoscope,
etc. The New Sydenham Society, London, 1861.

Anatomy of the Naso-pharyngeal Space.

The boundaries of the almost cubical space with which we
wish to become better acquainted by rhinoscopy, are in short as
follows: above, the body of the sphenoid bone; behind, the
basilar process of the occipital bone, and the first cervical verte-
bra; laterally, the pterygoid processes of the sphenoid bone and
the soft parts (the openings of the Eustachian tubes with their
prominences); in front, the posterior nares separated by the septum and bounded by the palatal bones, the inner lamellæ of the
pterygoid processes of the sphenoid bone, and the vomer;
within, the nares, the turbinated bones, and the *meati;* the inferior wall is wanting, and is temporarily supplied by the soft
palate, as it passes backwards and upwards in the movements of
swallowing, etc., and cuts off the naso-pharyngeal from the
bucco-pharyngeal cavity.

The accompanying wood-cut gives a sectional view of the

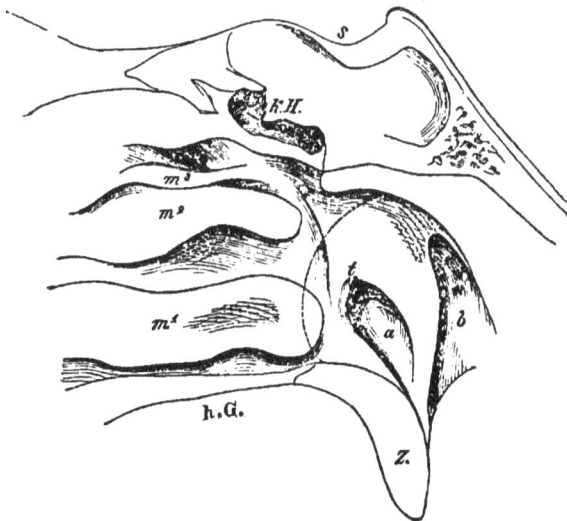

right half of the naso-pharyngeal space. In front are seen the
posterior extremities of the three turbinated bones, the superior

of which, m^3, runs backwards in a very narrow channel; the border of the septum is indicated by a dotted line, beneath which is the cut surface of the divided uvula, z. At about the height of the inferior turbinated bone, and posteriorly along the lateral wall, lies the pharyngeal mouth of the ear-trumpet, t. Below, and somewhat anterior to this, the velum, a, is attached, and from thence extends towards the middle line; it thus forms a swollen projection, which ascends towards the opening of the tube. In front of the opening of the tube, on the external edge of the nasal opening, a small shallow groove runs perpendicularly downwards, forming a slight projection. Behind the opening of the tube, somewhat obliquely downwards and backwards, there is a broad thick protrusion, a few lines in height, which springs forward into the open pharyngeal space, close to the posterior wall, forming, with the latter, a longitudinal chink-like cavity, the sinus of Rosenmüller, b. We shall hereafter have occasion to recur to the importance of this enlargement and of this sinus, in catheterization of the Eustachian tube.

The height of the posterior nares is, according to Moura, 12–38 millimètres (0·47–1·49 inches), the breadth, 25–30 millimètres (0·96–1·18 inches; the posterior wall of the naso-pharyngeal space may measure 50 millimètres (1·96 inches) in breadth and height; the height is, of course, measured from no fixed boundary below. The plane of the openings of the posterior nares is almost perpendicular. The diameter of the naso-pharyngeal space is about 20 millimètres (0·78 inches) from before backwards, and two or three times as great from side to side (Moura). The former diameter always diminishes downwards towards the uvula, and hence the indication for drawing the uvula and the velum forward.

Apparatus for Examination.

Czermak used at first in his examinations and demonstrations a common quadrangular laryngeal mirror, with rounded corners, and a bent, hook-shaped scoop, or sling, of wire, to lift up the velum and the uvula, and to draw them forward; afterwards he employed spoon-shaped and bent instruments, as the so-called palate-hook, or palate-spatula. Besides these, he used from the first, for the examination of a second person, the right-angled, jointed tongue-spatula of Petit and Simpson.

As regards the mirror, the author does not go so far as to recognise a special advantage in any particular form or substance. It need only be observed that, in general, for the examination of the naso-pharyngeal space, even under the most favorable circumstances, mirrors as large as those used in the examination of the larynx cannot be employed. When we are compelled from the relations of the space to use the smallest mirrors, the author prefers those made of metal, especially of steel, because they reflect more light in proportion to the surface than those of glass, in which also the metallic rim, although narrow, still diminishes the size of the reflecting surface. Still, the metallic mirrors are more sensitive and more difficult to retain in good condition; they rust easily, become scratched in cleaning, are easily discolored by warming, and are more difficult to restore. The size of the mirror which the author uses, is from 1–2 centimètres (0·39–0.78 inches) in diameter. The mirrors ought to be thick, in order that they may remain warm longer. For the inspection of the posterior nares, mirrors, in which the angle of attachment to the rod approaches nearly to a right angle, are the most advantageous. Voltolini also coincides with this view. Latterly, glass mirrors have been coated with quicksilver, and these excel in point of beauty and durability.

The mouth spatulas, made from Czermak's designs, can seldom be dispensed with in the examination of the naso-pharyngeal space; they have a spoon-shaped concavity in the blade intended for the tongue, are furrowed upon the lingual side, and are considerably bent longitudinally, so that the free end reaches far back towards the base of the tongue. Thus this spatula is also serviceable in laryngoscopy, since the tongue may, by it, not only be pressed downwards, but also forwards, by which movement the entrance to the larynx first becomes accessible to observation. It is well, therefore, generally to use the tongue-spatula in rhinoscopy, because it is only a question of gaining space from below upwards, of placing the tongue out of consideration.

The palate-hooks are made, according to Czermak's suggestion, of three different patterns, which are distinguished from each other by their breadth and curvature. The sling-shaped, fenestrated palate-hooks are specially intended for the examination of the posterior surface of the soft palate, but are otherwise less easy to handle, because the uvula easily falls through the fenestrum.

With these simple instruments the author has examined more than five hundred persons, and he feels no necessity for new contrivances, nor for further improvements of the apparatus for examination.

Illumination.

Experience has shown that we need a very good light for practising Rhinoscopy—more light, indeed, than for the examination of the larynx, a fact which may be explained by the use recommended of comparatively smaller mirrors. While for the examination of the epiglottis and of the entrance to the larynx, the divergent rays of a light room are sufficient when they are made to converge by a concave mirror, still in Rhinoscopy unfortunately not even a beginning can thus be made; it does not even answer for a repeated examination when the relations of the parts are already known from previous examinations. Sunlight can never be fully replaced by artificial light, and its chief advantage consists in its allowing the parts to appear in their natural color. But that consideration would be of little value with reference to parts, the natural color of which one has never seen; and a confusion in the determination of the colors can only take place when the examination has been conducted at one time with natural, and at another time with artificial light: for with the ophthalmoscope, the fact that the fundus of the eye is always shown in unnatural colors, disturbs no one. As we can seldom secure sunlight for rhinoscopic examinations, it would be better always to examine with artificial light alone, in order not to be obliged in individual cases to refer the colors found back to the actual ones. And sunlight, aside from the contingencies of the weather and the situation of the examining-room, can only then be used directly, when it enters comparatively horizontally, *i. e.* in the morning and evening hours, when the sun is low down. At other hours it may indeed be reflected in any given direction by means of a plane mirror, but the location of this mirror must be frequently changed, or recourse must be had to a heliostat. The author found it more convenient to throw a cone of light into the room by means of a concave mirror attached to the window, upon which the sun shone; then to seat himself and the patient within this cone, to collect anew the diverging rays by means of the illuminating apparatus, and then to apply them to the examination. Thus we obtain a soft light which is not blinding, and with which one might work by the hour together.

As for the artificial illumination, the light of any good lamp suffices, if it is made to converge by means of a concave mirror. Aside from the ordinary oil lamps, attempts have been made in various ways to secure a better light. Türck applied the electric and the Drummond lights; Voltolini passes a stream of oxygen through the flame of a photogen lamp, to secure a white and clearer light; his apparatus is described in the papers cited above. The solar lamps, which Störk employs, also give a very good light. The author has always been so well satisfied with the light of a middling-sized moderator lamp that he cannot yet become an advocate of the more elaborate methods of illumination. The objection which is made to artificial light, that it is in itself yellowish red, on the one hand fails for the above-mentioned reasons; and on the other, I may mention that Smith and Beck in London furnish a lamp-chimney of bluish-green glass (probably of cobalt) which also produces a perfectly white light, and which is equally well adapted to microscopic examinations.

Every artificial light for the purposes of examination must be first concentrated. For this purpose nearly every observer has made a contrivance of his own. Lewin, like Tobold, places over a lamp a kind of lantern, with a tube into which lenses are introduced. Stork has applied the so-called shoemakers' globes, i.e. glass globes of a considerable size, which are hung upon a frame before the lamp, and can be elevated or lowered at pleasure; these act as imperfect compound lenses, and the cone of light is allowed to fall immediately into the mouth of the patient; Türck, of Vienna, has, for some time, made use of his own illuminating apparatus, modelled upon the upper extremity of man, and admitting of a variety of motions. Mourá-Bourouillou has produced a contrivance, the so-called pharyngoscope, which consists of a concave or plane mirror of 15-20 centimètres in diameter, in the middle part of which an

opening of about 5 centimètres in diameter is cut, and into this
opening a lens of short focus is introduced. The flame is
placed behind this lens, and the rays of light fall converged
into the pharynx of the patient. The inventor states as a
special advantage of this apparatus, that with it one can shave
very comfortably!

Czermak and the author start with the idea of making the
illuminating apparatus fast to the observer's head, in such a
manner, that, when once it is arranged, it will leave both hands
free, and at the same time will permit one to follow slight
movements of the patient's head without interruption to the
examination. Czermak at first attempted to accomplish this
object by a modification of Cramer's forehead-band, and then
brought out his well-known mirror with a handle to be held in
the mouth. This is a concave mirror of 8–10 centimètres
(3·14–3 93 inches) diameter, and 20–30 centimètres (7·96–11·79
inches) focus, fastened into a fork which is attached to a metallic staff, and this last is fastened to a flat plate of wood or
caoutchouc (originally violet-root) in such a manner that the
observer can either hold this plate between the back teeth
of one side of his mouth, and thus have both hands free,
or he may place the plate in a straight line with the staff,
and thus use it as a handle. The author announced his
intention, while laryngoscopy was yet in its infancy, of com-
bining the illuminating mirror with spectacles. This idea
was pursued, and after many efforts resulted in the illuminating apparatus which, associated with his name, is now
largely exported from Vienna. In brief it is as follows: A
strong spectacle-frame has in the middle, on the bridge, a small
brass plate, upon which a second may be screwed. Both together form the shell of a small steel nut, which is fastened by a
small staff to the back of a glass concave mirror of 8–10 centimètres diameter, and 20–30 centimètres focus, not far from its
edge. At the middle point of the mirror the coating is removed
from a space about as large as a small lens, and through this
spot the observation is made. Moreover, the spectacle is so
arranged that lenses adapted to myopic or presbyopic observers
may be easily inserted in the settings. This mirror, by means
of the described nut-joint, is within certain limits movable in
three directions, and can be so turned as to be used by either

eye. The observer must look through the opening in the mirror, hence with that eye before which the mirror is placed ; for thus he looks exactly in the direction of the axis of the rays, that is to say, under the most favorable circumstances ; for precisely those parts which the mirror reveals will be most strongly illuminated.* Czermak draws attention to still other advantages of these contrivances besides the special one of the use of reflected light, and among them this, viz. the observer's own hand never interrupts his light, and he can examine the patient in any position necessity may require.

Care must always be taken not to burn the patient when using either converged sunlight or reflected light. It is only occasionally possible to see with both eyes, unless the mirror is placed upon the forehead, like Moura, or before the nose, like Bruns ; but in either case we lose the above-mentioned advantages of my arrangement. The successful use of this spectacle-mirror requires some practice, and, moreover, the individual mirrors are not all made exactly alike, so that one may be worn more comfortably than another. Thus the limited movement of the mirror and the prominence of the nose make it generally necessary to stand the light upon the left of the observer when he is to look with his right eye, and *vice versâ ;* this is especially true when the staff which unites the nut with the mirror is short, because then the mirror cannot be turned far towards the temple. The author has also found it serviceable, and in his practical courses has always given instructions, to place the light higher than the eye of the observer and the mouth of the individual to be examined, partly because the uncovered eye will not then be blinded with the direct rays from the lamp, and partly because in rhinoscopy one often has to use both hands before the mouth of the patient; and with a lower position of the lamp, it would easily happen that one hand might come between the lamp and the illuminating mirror, and thus cut off the light from the latter. When Voltolini urges as an objection to the spectacles, that the lamp must stand at the left when the right eye is to be used, he cannot seriously regard it as such, for the examination may be easily conducted thus. He draws

---

* The perforation of the mirror has been the source of much discussion in England ; the translator's experience has been that with the perforation a much clearer view is obtained.

attention, on the other hand, to one advantage of this apparatus, viz. that the observer can speak when the mirror is
placed upon the nose,—a matter of importance in the examination of aural patients.   Still further, the spectacles can
be very well applied to the examination of the external ear;
and the inventor has had the satisfaction of seeing them thus
used by the distinguished aurist Dr. Ignaz Gruber of Vienna,
as well as by Dr. Politzer, Instructor in Aural Surgery, also of
Vienna.   The spectacles are also adapted to the examination
of the larynx from the pharynx, or from the opening in the
canula, according to the experience of Czermak and Neudörffer,
and to the examination of the anterior nasal cavities.   Indeed,
if the author should divulge what he knows, he would say that
with these spectacles one can examine the fundus of the eye
very well, obtaining the direct image.

I may be still allowed to say a few words upon the use of
the spectacles.   They should sit so firmly upon the head as not
to slip nor to become displaced by its movements; to this end
the spectacle-bows must in some cases be first bent in accordance with peculiar conformations of the head.   It also happens
after long-continued use, that the joint becomes loose, and the
mirror is no longer firm: then either screw the plate up tighter,
or take it off, pass a file over it a couple of times, and then
screw it on.

When the lamp is placed near to, and behind the patient, the
observer with his spectacles must necessarily keep at a certain
distance from the patient, in order not to come into the shadow
of his own head.   The focus of the concave mirror is given in
figures, and is dependent upon rays, which, coming from an
infinite distance, may be considered as parallel, *i. e.* it is dependent upon sunlight.   If artificial light is used, the point of
union or crossing of the rays, *i. e.* the point of the inverted
image of the flame, is further from the mirror the nearer the
lamp is placed, and *vice versâ*.   But we cannot make use of the
image of the flame for illumination; the mirror, in order to be
equally illuminated, must receive the cone of rays before or
behind the image of the flame.   If we consider, moreover, the
distance of the lips from the posterior wall of the pharynx upon
which the light is to be directed, and the varying distance of
the point of clearest vision in different eyes, we obtain this rule

for illumination. The head of the observer should be at such a distance from the head of the patient, that the posterior wall of the patient's pharynx shall be at the point of clearest vision for the eye of the observer; and the lamp must be so placed that the crossing of the rays of the cone will fall immediately before or behind the wall of the patient's pharynx—the point where the mirror will be placed. The light forms upon the mouth of a patient a circular disk, in the middle of which is a round shadow corresponding to that part of the concave mirror which is not coated. Upon the pharyngeal mirror this shadow will not be seen. The concave mirror must stand as nearly straight as possible, in order that the opening for inspection may not be diminished, as it would be by an oblique position.

In order to use the spectacles with freedom, the observer must arrange them once for all before the commencement of the examination, so that he need not touch them again; but he must remember that the light will take another direction when he moves his head, and on the other hand that the patient's head is by no means nailed fast, but can be turned, elevated, or de-pressed, according to circumstances.

Türck, of Vienna, has lately become reconciled to the idea of attaching the illuminating mirror to the head of the examiner, and has proposed to construct an illuminating apparatus upon this principle, but of this nothing farther has as yet transpired.

The author goes by no means so far as to assert that illumi-nation can be alone produced by any given apparatus; on the contrary, that to which one has once become accustomed is for each the best. If, however, the widespread and constantly increasing use of the spectacle application is an index of its worth, the inventor may well believe that in a fortunate moment he produced an instrument of manifold uses.

Examination of the Naso-Pharyngeal Space.

Every patient who is to be examined for the first time, should first be instructed, in a few words, as to the method of procedure, and quieted, before the observer proceeds to the examination itself; still it is sometimes better, with very timid patients, not to delay, but rather to surprise them, as it were,

with the examination. The patient must be so placed that his
mouth shall be about as high as the eye of the observer. The
light is placed behind the patient at his side, and, indeed, at the
patient's left, when the observer wishes to look with his left eye;
the flame should also be somewhat higher than the observer's
eye (see above). Then the illumination will be satisfactorily
effected by giving a proper direction to the concave mirror.
Now the patient is requested to open his mouth wide, and to
breathe slowly and quietly, not too deeply, in order that the
velum may remain as quiet as possible. The point of the tongue
should lie forwards against the lower incisor teeth ; the tongue
itself should be flattened out as much as possible. Then the
tongue-spatula should be introduced, in order to see how the
patient will bear its pressure; at the same time the light falls
upon the uvula, and at its side as well as beneath it, upon the
posterior wall of the pharynx. The tongue-spatula must be
thrust far into the mouth, otherwise the posterior portion of the
tongue easily rises up and narrows the space. We consider it
important for the patient to keep his head erect, so that we may
look directly over the spatula, the lingual blade of which should
lie as nearly horizontal as possible, otherwise the knee of the
spatula casts a shadow in the mouth, or the blade of the spatula
gives a troublesome reflection.

Voltolini allows the patient to incline the head slightly forwards, in order that the flaccid velum and uvula may be drawn
down by its own weight, and removed from the posterior wall
of the pharynx, that thus the space, from in front backwards, may
be greater. Moura, on the contrary, desires the patient to bend
the head backwards, so that thus the plane of the posterior nares
should be inclined towards the mirror, that the inferior portions
might be more easily seen. Thus, the floor of the nasal cavities
slopes gently downwards and backwards, and forms, with the
horizontal plane of the spatula, an angle of 20° to 30°. The
reader observes that we take the middle course.

When the tongue-spatula has been applied, the palate-hook
must be introduced, in order to notice how the patient will bear
its use. The uvula, and a portion of the soft palate, is laid upon
the hook in such a manner that the uvula is directed forwards ;
then the hand is raised and the hook gently drawn towards the
operator until a slight resistance is felt, and, at the same time,

the soft palate is to be carried somewhat upwards by the hook.
It is well not to introduce the palate-hook in the middle line, as
then the hand which holds it is placed in front of the patient's
mouth. It is better to introduce it somewhat obliquely, so that
the handle of the instrument shall be about at the upper canine
teeth, and the operator's hand at the angle of the mouth. Thus
the uvula slips off less readily, and, with the fenestrated hook,
it does not fall so easily into the fenestrum.

If the manipulations with the tongue-spatula and the palate-
hook thus described, are well borne, the instruments should be
withdrawn and the patient should be allowed to rest, whilst the
observer warms the mirror—which is done most simply over a
flame; and indeed when artificial light is used, over the flame of
the lamp which serves for illumination. We prefer to warm
the reflecting surface, because this does not come in contact with
the patient, and the mirror therefore can be made a trifle warmer
without danger of burning the patient.* When the mirror has
the right temperature, which the observer recognises by proving
it upon his own hand or lip, the tongue-spatula is given to the
patient and he is directed to introduce it himself, while the
observer takes his mirror and hook and proceeds to the examination. Sometimes it is advisable for the physician to introduce
the spatula, and then give it to the patient to hold. Frequently
the patient presses the back part of the tongue too little down,
or slides the spatula too far out. This must be remedied before
the other instruments can be introduced. If the inner part of the
spatula lies too high, the right position for it can be given to the
patient by introducing the hook and pressing down the spatula.
It is advantageous to enter at the same moment with the mirror
and the hook, in order to lose no time. If the spatula is well
placed, and the palate-hook properly applied, let the mirror be
introduced, with its reflecting surface directed forwards and upwards, resting upon the spatula in the middle line, until the
posterior wall of the pharynx is reached. This is generally not
so very sensitive. At first one is amazed to see the distance

* Buzzard, in the *Lancet* for August, 1864, says that he has accomplished the
same object by spreading equally over the mirror a drop or two of a solution of glycerine and water, *partes æquales*; the mirror thus protected can be held for a longer
time in the mouth without being affected by the moisture of the breath, and the
sharpness of the image is not diminished. The translator has made a few examinations in this way, and with very satisfactory results.—(TR.)

over which the mirror must be shoved before it reaches the posterior wall of the pharynx. It is very easy to see the inner
extremity of the palate-hook. It is to be remembered that the
nasal septum stands above the palate-hook, the observer is
therefore obliged either to raise the handle of the mirror, or to
slide the mirror somewhat up along the posterior wall of the
pharynx, or to draw the hook more forcibly towards himself,
according to the given space, in order to see the nasal septum,
from which, as a clearly marked and established stand-point, we
proceed with the examination. If the nasal septum has been
recognised, a slight rolling of the mirror upon its stem towards
one side or the other, suffices to give us an image of the openings of the posterior nares and of the turbinated bones. If the
mirror is rolled a little further towards one side, we pass outwards
beyond the external wall of the opening of the nares; and if the
handle of the mirror is at the same time slightly raised, we gain
a view of the pharyngeal openings of the ear-tubes. The
author originally asserted that in examining the mouth of the
tube the mirror should be carried to the opposite side, and then
its surface, if it stands e. g. at the left, should be directed forwards, upwards, and towards the right, in order to gain a view
of the mouth of the left tube; but that procedure is rather complicated.

We obtain images of the openings of the nares in precisely
the same way, if the mirror, instead of being rolled sideways, is
slid towards either side, it being then placed opposite to these
openings. In order to see the inferior portions of the septum
and the openings of the nares, either a mirror very much bent
must be used, or the mirror must be pushed further up along
the posterior wall of the pharynx, until at all events its superior
edge is covered: thus the mirror has a more perpendicular position. For the examination of the upper and posterior wall of
the naso-pharyngeal space, we must employ mirrors which are
less bent, more like those which are commonly used for the
larynx. If the point, however, is to gain views of the posterior
surface of the soft palate, and of the palato-pharyngeal arch,
then the velum, according to its length and the amount of free
space which one has for the manipulation of the mirror, should
not be drawn forwards at all, or at least but very little, the uvula
merely being raised.

With regard to the facility of execution of this method, it may be observed—that the nasal septum, the openings of the posterior nares, and the posterior surface of the middle portion of the soft palate, are easily seen. The lower portion of the openings of the nares, the floor of the nasal cavity, and the inferior meatus, are nearly always concealed by the velum. The upper portions of the naso-pharyngeal space may also be easily seen ; the lateral portions are more difficult, including also the opening of the Eustachian tube; indeed, this last was not recognised for a long time ; the posterior wall is seen incompletely and much fore-shortened.

As for the position of the mirror, the following is almost of itself evident : the more perpendicular the mirror stands, the more clear images does it give of the anterior parts ; the more horizontal it stands, the more it illuminates and reveals the superior and posterior region. With an almost horizontal position of the mirror, we can gain images of the posterior nasal openings, but shortened, and hence unnatural. The higher the mirror can be introduced, and the more perpendicular it stands, the more natural and the more clear is the image which it gives of the anterior portions ; the anterior inferior portions (the inferior curve of the openings of the nares) can only be seen when the soft palate is drawn entirely forwards, or when it is very flaccid, and the space at the same time very large. The image in the mirror is reversed laterally ; that which is above is also in the reflected image above (and behind); what is below, is also below (and in front); the image of the right opening of the nares appears in the mirror at the left of the observer, and that of the left at his right. It should be mentioned, moreover, that such reflected images never give with entire accuracy the true size of the object. A single consideration will convince every observer why it is so. The reflected image could only correspond exactly to the size of the object, when the corresponding axes of the object and the mirror stand in a right line with each other.

Even with a respectable amount of anatomical knowledge, the majority of practising physicians will hardly be sufficiently clear upon the details of the naso-pharyngeal space, to be able to forego preliminary exercises upon the cadaver, and upon good models.

The Rhinoscopic Image.

It is clear from what has preceded, that an image of the naso-pharyngeal space can never be gained from *one* position of the mirror, but only from numerous turnings and slidings of the mirror, and even from the application of two mirrors bent at different angles. Hence it is manifest that these images, like all others in fact, have only a relative degree of truth. This is more true of drawings and colored representations even than of the reflected images in the mirror, for in the former the peculiar difficulty of coloring is added. But every one has agreed to disregard this, otherwise it would be utterly impossible to offer illustrations of the appearance of these parts. There is still to be mentioned the peculiar circumstance that the walls of this more or less cubical space, which occupies us, appear in the image to be spread apart from each other, and hence individual parts are brought into such relative positions, that it will at first be rather difficult for the observer to feel himself at home.

We invite the reader to look at the first figure in our first plate. It gives the normal appearance of the naso-pharyngeal space, and was made after numerous examinations by sunlight upon the cadaver and upon the living subject. At the same time we have represented Itard's catheter, introduced on one side, the left in the figure, into the right sinus of Rosenmüller, and on the other side, the right in the figure, into the opening of the left Eustachian tube.

In the middle of this figure, standing upright like a column, is seen the nasal septum, as narrow as a band at its lower end, and of a pale yellow color; above, where it becomes broader, it is red like the mucous membrane of the mouth and pharynx. The openings of the nares appear sharply defined; below, on each side, there is an imperfect image of the lower turbinated bone; above it, comes the middle turbinated bone like a round knob, directed inwards and downwards, partly covered by the preceding; and finally, the superior turbinated bone brightly illuminated like a short, narrow band. Between the turbinated bones and the septum is the free area of the cavity of the nose; between the middle and inferior turbinated bones, towards the outer edge of the nasal opening, is the middle meatus in shadow; and above the middle turbinated bone is the superior meatus. At the same time it is seen that a fourth meatus exists, in fact, almost always above the superior turbinated bone. If the mirror is placed far up and perpendicularly, or if individual parts are wanting, we can see farther into the nasal cavity, especially along the side of the middle turbinated bone. Thus occasionally our vision may penetrate even to the mucous coating of the nasal bones, and of the cribriform plate of the ethmoid bone. If the septum stands to one side, or if the mirror is placed somewhat obliquely, then we look in, some distance along the septum, and either one of the openings of the nares may seem narrower than the other. This impression is corrected if the mirror is properly placed and introduced with the other hand.

Below, the image of the nares is cut off by a curved red stripe, the reflection of the everted velum, the border of which, on both sides, rises up over the nasal openings, and then bending round, bounds, by a very prominent swelling, the entrance to the Eustachian tube. Between the latter and the nasal opening there is seen a pale reddish-yellow surface, which is bounded externally by a projecting red swelling in the angle; this projection and this pale surface are important points in finding the mouth of the tube. Externally from the nasal opening a longitudinal channel descends perpendicularly. (See Anatomy, page 5.) It will be observed, that the pharyngeal mouth of the Eustachian tube appears to lie with its projection close to the nasal opening, while in point of fact it lies to the side and behind this opening. The projection in the figure lying beneath the mouth of the tube belongs to the velum, and lies in the living subject below the mouth of the tube; the third projection lying above the mouth of the tube, is in fact behind it, and the sinus shown between the nasal openings and the mouth of the tube, is really perpendicular, and in front of the latter. The red uneven surface above the nasal openings is the representation of the upper wall of the naso-pharyngeal space, which passes without any sharply defined boundary into that of the posterior wall. The shadow running upwards and outwards above the projection over the mouth of the tube, indicates the sinus of Rosenmüller, lying between the projection and the posterior wall. Thus we see that the posterior wall seems not only fore-short-

ened, but also much broader than it really is. The color of
the mucous membrane of the nasal cavity and of the turbinated
bones is in the image a pale reddish-yellow, or a greyish or
bluish-red; the broad surface at the opening of the tube is a
pale reddish-yellow; the rest of the mucous membrane upon
the upper portions of the septum, upon the velum, upon the
posterior projection of the mouth of the tube, and upon the
upper and posterior wall, is bright red, like the mucous membrane of the pharynx and of the mouth. We observe here,
that in the course of a long sitting the septum and the velum
become manifestly much redder during the examination.

We have seen the image reflected in this manner many hundred times, and so have all those also learned to see it who have
studied with us, and thus too have we demonstrated it often
enough. We cannot forbear noticing in a few words the differences which exist between our figures and the others which have
hitherto appeared; among them are those of Czermak, Voltolini,
Merkel, and Wagner.

Czermak's figure in his *brochure* is manifestly only a sketch.
Voltolini's figure makes more pretensions, and is also better
drawn, but in it the turbinated bones can scarcely be found.
Merkel, of Leipsig, also gives in his work cited above, a rhinoscopic picture; but this represents by no means a physiological
condition. He says, "I hope that this representation will credit-
ably distinguish itself from all the anatominal illustrations
•hitherto made of this region of the body (not even excepting
those of Czermak and Semeleder)." It does in truth distinguish
itself greatly, and it is so remarkable, that the condition it represents deserves to be received into our collection of rhinoscopic
rarities. The best representation hitherto is that of Dr. Wagner of New York, which also shows the entire septum, and a
portion of the floor of the nasal cavity, and which is drawn
from observations upon himself; but the region of the mouths
of the tubes is still not clearly rendered.

Latterly Czermak,* the facile experimenter and the unweary-
ing investigator, has, in pursuing self-demonstration, produced
photographs of the larynx and of the nasal openings, and of the
former, even stereoscopic views. It is known that many years

---

* Report of the Session of the Vienna Academy, Nov. 7, 1861.—Comptes-ren-
dus, 25 Nov. 1861.

since futile attempts were made in this direction. If we accord to these no special practical value in consequence of the great delicacy and difficulty of the task, yet they give us another opportunity of admiring Czermak's wonderful facility, in the manipulation of self-demonstration and also his untiring perseverance.[*]

Difficulties.—Peculiar Cases.—Expedients.

Rhinoscopic examination, in comparison with laryngoscopic, succeeds readily, and on the first trial, with a very much smaller number of persons. Even now, when so great a number of examinations have been made, we cannot determine upon a comparative percentage; for here the practice and the adroitness of the observer are still more important than on the examination of the larynx. We can tell quite well by looking at any patient, whether he will be easily examined with the rhinoscope or not. If, when the patient opens his mouth wide and puts out his tongue, the entire uvula hangs freely down, the examination will, in all probability, be very easy, as the relations of space are favorable. The difficulties which oppose themselves to rhinoscopic examinations are less numerous than in laryngoscopy, but in turn they are less easily removed. These difficulties, setting aside, of course, the possible adhesions of the soft palate, or its deformities, arise only from the tongue and from the velum.

But few persons can, without long practice, hold their tongues so low, and so quietly, that we can forego the use of the spatula. Many unconsciously press the tongue so constantly and strongly upwards, that we are scarcely able to hold it down with the spatula. With persons who hold their tongue very unquietly, there are the following additional disadvantages: when the spatula is forcibly pressed down upon the tongue, the latter suddenly yields, the spatula slides in a little further, and, by touching the base of the tongue, excites increased choking;

---

* The translator had the pleasure of seeing these pictures at Prof. Czermak's house before they were published, and he cannot but express his amazement at the success attending so difficult an undertaking: nor can he fail to mention the great satisfaction he received from the demonstrations which the honored Professor was kind enough to make for his benefit.

or the tongue moves constantly up and down, rubs itself upon
the end of the spatula, is thus tickled, and it is extended and
withdrawn by turns; or from the narrowness of the space the
mirror must be laid upon the tongue, and the tickling produced
excites at the base of the tongue the unfavorable movements
mentioned. All this may happen, whether the physician or the
patient holds the spatula. Generally, as we have also observed
in practising laryngoscopy, the base of the tongue on an
average is more susceptible to the touch, than either the velum
or the posterior wall of the pharynx. And yet the pressing
down of the tongue can only very rarely be avoided.

The velum offers a whole row of difficulties. It is the first
necessity of the examination, that the velum should hang down
flaccid; if it is drawn up it approaches at once the posterior
wall of the pharynx, diminishes the space through which we
have to look, and covers the parts to be examined. If in its
contraction it reaches the posterior wall of the pharynx, the
naso-pharyngeal space is completely shut off from the bucco-
pharyngeal cavity, and we only see in the mirror the anterior,
now inferior, surface of the soft palate. There are individuals
who, when they open the mouth, draw up also the velum;
others draw it up even when the instruments are brought near
to their mouth, or when the velum is ever so lightly touched;
and indeed the strongest exercise of the patient's will has here
no effect. Frequently the hook, if it has been applied, will also
be drawn up, and pressed against the posterior wall of the
pharynx. How can we overcome these obstacles? A forcible
pull with the hook avails nothing, even when this instrument
can be introduced. Other instruments have therefore been devised, in order to secure the application of more force. Türck
has produced a number of them. First, a kind of polyp-forceps,
with which the uvula was to be seized, and the soft palate drawn
forwards; then a forceps like Civiale's modification of Hunter's
urethra-forceps, the separating spring blades of which are closed
by sliding the outer tube, thus clamping firmly the uvula; then
for the same purpose an instrument resembling a lithotrite, the
arms of which are opened and closed by sliding, and may be
fixed by a permanent screw. This same learned inquirer also
applied to the curved end of the palate-hook a horizontal gutter
in order to prevent the velum from slipping off, and by which

he might be able to draw more firmly. He recommended a small tube with a wire sling, in which the uvula might be caught and held, and perfected its application in the so-called palate-depressor, which, with a sharper curve than the palate-hook, is intended to draw the soft palate forwards and downwards, the instrument being fastened to the forehead by a string, and the uvula-sling being applied: he finally contrived an instrument, modelled after the style of a lithotrite, or after Brambilla's œsophagus-forceps, which is to be placed behind the soft palate, with one arm resting upon the posterior wall of the pharynx; by drawing back the other arm it is opened, and the velum is in this way to be pressed forwards. Voltolini speaks most decidedly against these instruments, and Störk makes the biting remark, that by the use of them the pharyngeal space may certainly be enlarged, but "permanently" so. So Markel says, these modifications, following so quickly one after the other, could not have spoken favorably for the various appliances themselves, which produced partly coughing, partly extravasation of blood, partly swelling of the uvula, and could not accomplish their object, for *we cannot by force pull forward the elevated velum.* The proposal of Auzias-Turenne to anæsthetize the velum with bromide of potassium, has not proved of any avail.

What, then, is left for us to do? The point is to render the velum flaccid; then it is of itself removed from the pharyngeal wall. Without discussing the point farther, as the patient cannot perform nor understand this procedure, it had been thought best to accomplish it by strategy. The author hoped to accomplish the purpose by directing the patient to breathe through the nose instead of through the mouth, for then he must allow the velum to hang flaccid; but even this is not so easy. Czermak gave a better suggestion in causing the patient to utter well-marked nasal sounds, for in this way, too, must the velum become flaccid. Still the result is frequently not reached, inasmuch as the patients utter no nasal tones. Voltolini thinks that pressing the tongue down forcibly would suffice to make the velum flaccid, or at all events when it was relaxed to retain it so; this observation, however, has not been verified by the author.

In such cases we must renounce the idea of proceeding rapidly, and determine by repeated touching and rubbing of the velum to diminish its sensibility to such a degree that the examination may be carried out; or we must determine to make the examination without the aid of the palate-hook. Such unfavorable cases frequently occur exactly then, when the patients intend to do right well, and think constantly of the examination; to hold quietly the mirror in the pharynx, and by speaking of other things to draw off the patient's attention from the operation, sometimes succeeds, and the velum drops down; or if we weary out the patient, the result is the same. But sometimes, especially when respiration is interrupted, a sort of tonic spasm of the muscles of chewing and of the pharynx occurs, and the author has occasionally seen the saliva jet forth in a stream from both the ducts of Steno into the cavity of the mouth in consequence of the muscular pressure.

One error which is often made in the use of the hook is, that the operator pushes the velum upwards and backwards instead of forwards, in consequence of which he fills up the space; or he may turn the velum forwards while he does not draw upon the hook, but shoves it in, and consequently the surface of the hook covers the parts to be examined. In many cases, as Voltolini also observes, we find the naso-pharyngeal space filled with large mucous bubbles, which by their reflection render the examination impossible. The simplest way of removing these bubbles is to blow into the patient's mouth; with intelligent patients Voltolini makes use of a camel's-hair brush for this purpose.

Sometimes when the velum is very long, and the tongue cannot be well pressed down, there is not room enough to bring the mirror into its proper position; in such a case it is well to introduce the mirror with its reflecting surface downwards, until it is behind the tongue, and then first to turn it upwards.

Not unfrequently when the examination lasts too long, coughing follows. This occurs because the saliva collects at the bottom of the bucco-pharyngeal cavity, and in running down arrives at the entrance to the larynx. Hence it is well to let the patient swallow immediately before the introduction of the tongue-spatula.

Loss of substance on the soft palate does not always facilitate the examination, for if the arches of the palate are destroyed, the velum stands generally more horizontally, in consequence of the loss of the natural tension downwards. In perforation of the soft or hard palate, we can occasionally pass small mirrors through the opening (Gerhardt). In cleft palate the examination succeeds very easily, but it is not very instructive, because the relations are entirely altered, especially when the pressure is continued through the hard palate. The opening of the tube also becomes entirely different, as the lower projection has a different direction. The difficulties growing out of partial adhesions of the soft palate are, as might naturally be supposed, different in each case, and not to be defined.

If one was to conduct the examination in the manner described above, with the tongue-spatula, the palate-hook, and the mirror, he would need three hands; hence the tongue-spatula is committed to the patient; the mirror or the palate-hook he cannot, of course, introduce. If we would still further undertake therapeutic or operative manipulations, another hand still would be necessary. But notwithstanding this necessity, man is created with but two hands. The patient, as has been said, can only take the tongue-spatula; assistants are not always at hand; and if four hands are employed in front of the patient's mouth, the space is narrowed altogether too much. An assistant, too, is of very little use, as he can only see at the same time with great difficulty, and yet this would be necessary. These inconveniences have led to the invention of various instruments, the purpose of which is to combine two or even all three of the necessary instruments, that they may be introduced with one hand, and that the other hand may be free.

Rhinoscopes.

The one of Czermak's which was brought forward as an experiment, consists of two short tubes, one of which slides in the other; the interior is prolonged into a semi-tube, which bears externally a ring for holding it firmly; the external one slides along the preceding by a rod, which has a hook for seizing hold, and which runs along the half-tube of the interior piece. The outer portion carries, above, a perpendicular little plate; on the interior portion, the extremity from above downwards is cut off obliquely, and in the opposite direction to this cut surface, a mirror is fastened at an angle of 45° to the longitudinal axis of the tube. In this instrument, the outer tube, with its little plate, represents the palate-hook; the interior one represents the mirror, and, with its half-tube, the tongue-spatula. The instrument is closed when the tubes are completely slipped into each other; it is thus introduced, and then opened by a sliding movement, in consequence of which the velum is brought to lie upon the external tube.

Störk's rhinoscope is, in a few words, put together as follows: Suppose a mirror and a hook, both having rings on their extremities; their handles, at the middle point of their stems, are united like a pair of scissors, so that by pressing together the separated rings, the hook lying upon the mirror is raised; the mirror, together with its stem, has a longitudinal sliding movement; whilst, upon the anterior portion of the mirror stem, a small leaf may be placed as a substitute for the tongue-spatula. The mirror and the hook are pressed against each other by a spring. The instrument thus closed, is introduced behind the velum, and then, by an approximation of the rings, opened, and the hook, upon which the velum now rests, is drawn towards the observer. Neither of these instruments is altogether handy and convenient. The author, by the greater facility acquired in the use of simple instruments, has become convinced that the day of the rhinoscope is past.

Voltolini's instrument is a long, black spatula, on the end of which, at a prescribed angle, a steel mirror with a short neck is placed: the advantage, however, is not great, inasmuch as a second hand must still be used for the application of the hook; and, moreover, as a rule, the patient himself can hold the spatula.

If the palate-hook is not easily submitted to, let the examiner take the tongue-spatula with one hand, as he can handle it to better purpose than the patient, and with the other hand let him then introduce the mirror, and make the examination without the hook; so also, if the patient cannot himself apply the tongue-spatula.

If the point is to perform a surgical operation by the aid of the mirror, "to make the eye the guide to the operating hand" (Czermak), then the surgeon must hold the mirror with one hand and guide his instrument with the other, and the patient must himself press down his tongue. Voltolini uses his spatula-mirror in such a case, or he slides, upon the stem of an ordinary mirror, plates of gutta-percha, of various sizes, for tongue-spatulas. Störk has for the same purpose mirrors, the stems of which have wing-like attachments. The author would also make use of the same instrument if the patient could not press down his tongue himself. In all these cases we must renounce the use of the hook, as neither hand is free for its use, or we must contrive some other way to elevate and draw forward the velum. Türck has proposed to introduce the hook, and to hold it fast by means of Kramer's forehead-band and a pincette, or to apply in the same way the uvula-holder, or the palate-depressor and uvula-sling, and, by means of the strings attached to them, to make both fast to a band passing over the forehead. But we would recommend none of these contrivances, for, if the tension is annoying to the patient, he will become so restless that nothing farther can be done. The first proposition would be the more feasible, but still more simple is the idea of Wagner. He twists a wire into an instrument shaped like a velum-hook, and curves it so as to receive the uvula, whilst the free ends of the wire having been made a spring, by being twisted together, are curved upwards, and, by their elastic force, can hold firmly on the nose like a clamp. If the pressure is too strong upon the velum, it can easily slide over the hook without the patient suffering any injury, and we must then commence over again. Wagner's idea of inducing his patients, by the use of his hook, to paint or cauterize themselves, shows great disinterestedness. Czermak has made verbally the very important suggestion of introducing in such a case, through the nasal cavity, instruments like the canula of Bellocque, so that the strong watch-spring curving forward into the pharynx would keep the velum pressed forwards and upwards.

Examination without the Hook.

The above-mentioned method of examination without a hook depends upon the following: Between the uvula and velum there remains on either side a free arched space, through which a small mirror can be introduced without touching any portion of the structure; and from this same point, with a little pains, almost the whole naso-pharyngeal cavity can be examined; and still farther the mirror can also, without coming in contact with the parts, be directed towards the other side, by passing the stem beneath, and the mirror behind the uvula, thus making the examination complete. Thus the operator always has one hand free when the patient himself presses his tongue down. This mode of examination has been practised by the author for a year and a half, and has also recently been described. It is much less wearisome for the patient than any other; for the examiner it is troublesome, as small mirrors alone can be used, and only small sections of the space can be seen at one time; it requires also a very exact knowledge of the parts to be examined, and is in no wise adapted to beginners. Its application, however, is dependent also upon the fact that the tongue can be pressed down, and that the velum hangs loosely. The author cannot allow the general importance which Voltolini attaches to this method. For the examination of the pharyngeal openings of the Eustachian tubes it will generally answer, as they are situated laterally, and a diminution of the space from before backwards is of less consideration; but in order to give a sufficiently distinct view of the posterior nares and of the septum, the hook is, except in very rare cases, exceedingly useful, if not indispensable: for without it, frequently the upper edge only of the posterior nares is seen, as indicated in Figure 4, Plate II., because the velum, even if loose, does not stand far enough off from the posterior wall of the pharynx. With this method the author has once experienced at an earlier day upon himself, and frequently observed in his courses, the following appearances: A circular boundary was seen, and within, as if set in a frame, a knobbed, swollen, reddish body, being the arch of the mucous membrane between the uvula and the velum. The socket of the posterior molar tooth, together with the tooth, was also visible. The observer supposed he had before him the edge of the posterior nares, and a turbinated bone either pathologically changed or covered with mucus. At other times the posterior surface of the velo-palatine arch is seen on the triangular fossa above the tonsils, between the two palatine arches. One can always see something, but at first and for a long time he does not know what he does see, and hence he cannot give the necessary alteration to the position of his mirror.

Magnifying Apparatus.

So long ago as 1859, Dr. Wertheim of Vienna recommended the use of small concave mirrors in order to obtain magnified images. Türck makes the objection to these mirrors, that in laryngoscopy, in the examination of parts which are situated at different distances, we must use mirrors of different foci, and of so much the greater focus the farther the objects are from the mirror: he also thinks they have but small magnifying power. He has himself brought forward for the same purpose a simple Galilean telescope, which can be made fast to one of the arms of his illuminating apparatus. Voltolini had an opera-glass from which he had removed the ocular for magnifying the image; and also a small telescope, from which he took all the glasses except the objective; he obtained good results when he had the patient's head held firmly by a special contrivance; still this was achieved by sunlight alone, and he rightly thinks that for such peculiar examinations one would have time enough to wait for sunlight. The author may be allowed to remark with respect to the small concave mirrors, that for the examination of the naso-pharyngeal space one mirror should suffice, as all the parts to be examined lie at almost the same distance, and that but slight; but concave mirrors only represent the objects enlarged in the centre, while on the sides they are distorted. And then the author cannot allow the necessity or even the usefulness of such apparatus: for alterations so slight as not to be detected with the naked eye, or with the aid of spectacles, can hardly give occasion to an examination, or cause discomfort.

Measuring Apparatus.

It has been thought desirable by many to apply a scale to the mirror, in order to be able to give exactly the size of objects seen. Mandl in Paris scratched such a scale upon the surface of the mirror itself, but by this method too much of the reflecting surface was lost. If something of the kind is desired, the author would recommend that the scale should be marked upon the rim of the mirror, but he considers all such contrivances as unnecessary, inasmuch as we always have so simple a method of determining the relative size by a comparison of the magnitude of the image with objects generally known. Why the size of the image always deviates a little from the size of the object, has been indicated above.

Double Mirrors.

Czermak has suggested that by the use of double mirrors we can obtain views of the floor of the nasal cavity, which in the above indicated method of examination remains almost always in shadow; thus the light is thrown from the first mirror upon the second, and from this upon the object; it then pursues the same path back again, so that the image of the second mirror reproduces itself in the first. He has recommended a similar procedure for the examination of the laryngeal surface of the epiglottis. These double mirrors have hitherto been tried only by Dr. Wagner and by Voltolini. Wagner says of them: "The double mirror is just as Prof. Czermak describes it: the upper, together with its image, is reflected in the lower. In the examination it comes to stand in such a way that the upper reaches into the posterior nares. As the introduction after the second or third attempt succeeded so well that the mirror was no longer tarnished with mucus, I thought that I should now have a fine view, but I saw that I was deceived: for in spite of the most skilfully conducted attempts, and in spite of the bright sunlight which I used for the illumination, the upper mirror constantly gave me only the reversed image of the lower, viz. a portion of the posterior surface of the soft palate, and a part of the palate-hook. This want of success had its foundation only in this, that the floor of the nasal cavity did not receive light enough by the illumination through the mouth to give forth a reflected image. I therefore sought to procure more light in some other way, and while the pharynx and mouth were not illumined, I threw in light by means of an illuminating mirror through the nostrils. The result, as might have been expected, was good, for I saw a portion of the floor of the nasal cavity clearly illuminated. This entire method of experimenting, however, is very laborious; the illumination does not always succeed, and the image which is obtained includes but little. I think, therefore, after my experiments, I am warranted in saying of the double mirrors devised by Prof. Czermak, that by means of them, and by illumination through the mouth, it would not be possible to obtain an image of the floor of the nasal cavities, and that this result can only be reached by introducing light into the nasal cavities through the nostrils, an experiment which costs much labor, and which is not proportionately rewarded by the diminutive image obtained." The author would only allow himself the comment, that it was, perhaps, still possible to carry out Czermak's proposal if the two mirrors were properly inclined towards one another; but there remains always something uncertain, if we receive at one time with the single mirror a singly inverted image, and at another with the double mirror a doubly inverted image, and then have to reduce both images to each other and to the natural scale. Therein lies the difficulty of the procedure from a theoretical point of view.

The last mentioned process of Wagner's naturally leads to the

Anterior Examination of the Nasal Cavity,

which presents a necessary supplement to the rhinoscopic examination, and should not be neglected in any pathological case. The illumination must be made by some one of the described methods. If, then, the point of the nose is pressed upwards and the alæ nasi outwards, we see quite far back in the nasal cavity the anterior end of the inferior turbinated bone, portions of the middle one, of the floor of the cavity, and of the septum. This is accomplished still better, if we apply, for the dilation of the nostril, a two-leaved ear speculum, or the nasal speculum of Markusovski of Pesth, by which also the bristling hairs, which otherwise produce disturbing shadows, are pressed aside. By means of the last instrument we can sometimes see even the posterior wall of the pharynx ; and Czermak once happened to see on the cadaver, with a small mirror, a portion even of the inferior surface of the inferior turbinated bone, and the openings of the nasal ducts. Voltolini has devised still another method. He slides a strong, well polished ear catheter, or a bright straight tube into the nasal cavity, and thus examines anteriorly to the best advantage, with sunlight. The catheter, when it is moved, illumines the surrounding parts by reflected light. Voltolini could thus see as far as the posterior wall of the pharynx, and the swelling at the mouth of the Eustachian tube; or he introduces a catheter, as described above, and examines through the other nostril. The catheter then illuminates by reflection when it is moved, and its reflex can be accurately followed. But Voltolini himself adds that this is only the exception, and that in particular the left nostril is but very slightly adapted to it, for the nasal septum generally inclines towards one side, mostly towards the left, in consequence of which the cavity is often very much narrowed, so much so, that one has difficulty even in introducing the catheter; and aside from these deviations, the nasal septum exhibits frequently large outgrowths with sharp corners, as is seen in two preparations in the author's possession, one of which is of the size and almost of the shape of the middle turbinated bone, whilst the septum stands perfectly straight, and the turbinated bones of the corresponding side are uncommonly small. These portions are scarcely ever regarded at post-mortem examinations.

Illumination by Transmission.

This procedure originally advanced by Czermak for the larynx, and then also applied to rhinoscopy, consists essentially of the following: The mirror itself is introduced into the dark pharyngeal space; the parts to be examined, the larynx or the nasal cavities, are illuminated by clear concentrated light falling through from without, and giving in the mirror a glowing red reflection, shaded off, which in the case of the nasal cavities scarcely allows us to sufficiently distinguish the peculiarities of form. In this respect the transmission of light presents no practical significance for rhinoscopy.

Demonstration upon One's Self and upon Others.

Self-observation must be regarded as an actual aid in arriving at an exact knowledge of the naso-pharyngeal space, and in recognising what one may attribute to the parts, and what not. Self-demonstration, which is not to be separated from self-observation, is no less indispensable, on the one hand, to convince the still large number of the incredulous, and on the other to secure for those, who for the first time would engage in the study, a sufficient acquaintance with their field of observation—a fundamental necessity for rapid progress. For self-observation, and particularly for demonstration, according to Czermak's experience, artificial light is the best, and so also the apparatus devised by him, which has long since become widely known by his travels (see also his pamphlet on the laryngoscope, etc., cited above). The plane mirror, attached to the apparatus, must stand rather more perpendicularly than in the examination of the larynx. The demonstrator must be able to hold his tongue quietly without spatula or cloth, and to press it down, as he needs both hands to introduce the palate-hook and the mirror. A quite perfect view may be attained, but ordinarily a long time elapses before one is at home in the proper management of the two mirrors, of his tongue, and of his velum.

Voltolini says in his last larger work, p. 21, that it is only in self-examination that he is obliged to elevate the velum with the hook or with Störk's rhinoscope, and he explains the circumstance from the fact that the eye stands higher than the uvula, and hence one cannot look under his own uvula. But the reason does not lie in this, for the author has many times made observations and demonstrations upon himself without the hook, and especially in each of his courses; and he employs this method of demonstration when he introduces Itard's catheter, in order to turn it with the free hand and place it in the pharyngeal opening of the Eustachian tube.

In connexion with this there is another kind of demonstration, viz. that upon the sick. Very frequently, when one makes an examination before other physicians, he is asked to show what he finds; it may be either from curiosity or from a want of confidence. It is, indeed, difficult to satisfy this demand in laryngoscopy, but in rhinoscopy it is much more difficult. At the examination, as has been described in the preceding sections, there is scarcely room near the head of the observer to look into the pharynx of the patient. This becomes possible only when the individual removes his head farther from the patient, so that the pharyngeal mirror first receives the rays of light beyond their point of crossing. It is accomplished much better, however, when the second observer places his eye close behind and near to the illuminating mirror, and looks past this upon the mirror in the pharynx. Still further, the following must be borne in mind: when I am examining, and place the mirror so as to reflect exactly the septum, the second observer, who is standing close at my right, sees the external edge of the image of the right nasal cavity, towards the mouth of the tube; if I wish to place the mirror so that the second observer shall see the septum, then I cannot see it at the same time with him, but I shall have the image of the left nasal cavity, etc. The reason of this is explained by the consideration of a simple physical law, from which we will excuse the reader.

Hitherto we have only spoken of introducing the mirror with one hand; those peculiar cases which require the mirror to be introduced with the right or left hand alone, will be spoken of hereafter. When in pathological cases the point is accurately to determine the existing condition, it would be well to introduce the mirror alternately with one hand and then with the other.

CHAPTER II.

Practical Application of Rhinoscopy.

WE shall endeavor in this division, by the citation of observed cases and of actual results, to establish the worth of the rhinoscopic method brought forward by Czermak for practical medicine, and thus to secure for this branch of science the consideration which it deserves, but with the permission of the reader, we will first bring especially to his notice another subject, which certainly is also not without interest. We refer to

The Catheterization of the Eustachian Tube and its Relation to Rhinoscopy.

The treatment of the diseases of the ear forms, as is well known, one of the weak points in the healing art, in consequence of the concealed position of the organ, which renders it only in a limited degree accessible for diagnosis as well as for therapeutic applications; only the external ear, and perhaps the tympanum, in case of the destruction of the membrana tympani, are accessible to the eye and to the hand of the operator. The Eustachian tube, indeed, falls still less within the domain of a physician's powers; for its mouth, situated at the base of the cranium, deep in the upper part of the cavity of the pharynx, is perfectly withdrawn from ordinary inspection, and almost entirely, also, from the touch of a finger introduced into the mouth. The only means of bringing the Eustachian tube within the limits of the physician's domain has been, and is still, the introduction of a catheter through the mouth, or the nasal cavity. That this procedure has its peculiar difficulties, only those would dispute who have not often attempted to pass the catheter into the mouth of the tube. The prescribed rules, exhausting as they do so completely what might be said upon the subject, amount in fact to the following : bring the catheter to the right spot, then make a correct and skilful turn or sliding movement with it !

Let us regard the chief points of the procedure a little more closely. In all cases the prominences in the vicinity of the mouth of the tube, described under the head of anatomy, play an important part, for the peculiar feeling which is imparted to the hand when the catheter, jumping or sliding over one of these prominences, enters into the mouth of the tube, is a sure ground for recognising the success of the operation. Kramer passes the catheter in, until it touches the posterior wall of the pharynx, and then draws it slowly backwards, while at the same time he turns the beak of the instrument outwards and upwards ; thus the small head of the catheter passes over the posterior lip of the tubal prominence, and slides out of the sinus of Rosenmüller into the mouth of the Eustachian tube, after which the catheter is passed still further into the tube in the direction of its curvature. Somewhat more clear and more sure does this proceeding become, if the catheter, when in contact with the posterior wall of the pharynx, is turned a little more than a quarter of a circle outwards and upwards ; thereby the head of the instrument enters into the sinus of Rosenmüller, whilst the direction of the beak is recognised by the ring on the external end of the instrument. If the instrument is then gently drawn straight out of the nose, the beak slides over the posterior prominence into the mouth of the tube with a perceptible jerk.

Another method is the following : the catheter is introduced as ordinarily through the inferior meatus, with the point slightly directed towards the septum, in order not to catch upon the turbinated bones. When the beak of the instrument stands free in the cavity of the pharynx, the external end is raised and the beak turned downwards ; thus the curvature lays itself upon the upper surface of the velum. Then the beak is turned outwards and upwards, and moved slightly forwards, by means of which, passing over the ridge in front of the mouth, it enters the tube.

Many recommend that the patient should be told to swallow at the instant that this manœuvre is made, and flatter themselves that by the elevation of the velum the beak of the catheter will

be of itself, so to speak, thrown into the mouth of the tube. This
motion of swallowing follows as a rule, even when it has not been
specially ordered.

If the catheter is once within the mouth of the tube, the curvature of the beak at every elevation of the velum, and also in
swallowing, is pressed upwards, and thus the instrument is rolled
out of the tube. This is especially true of the heavier metallic
catheters.

The position of the mouth of the tube, when the velum is
elevated, is not without interest. The latter rises gradually so
high that it touches the posterior prominence, and finally covers
the mouth of the tube completely—see Fig. 4, Plate II., repre-
senting the mouth of the right tube. The author has observed and demonstrated these two conditions upon himself.
Dr. Pollitzer notices these observations, and in a communication
upon "the relation of the Trigeminus to the Eustachian tube"
says, of the position of the mouth of the tube in swallowing,
"the levator palati mollis will raise the velum when it is flaccid. Having now in mind its fixed point (the cartilage and the
lower membranous edge of the tube), we find that even Tortual
disputes the opinion, that this muscle, by swelling out when
it contracts, must cause a constriction of the tube, which Tortual
represents as always gaping. This view, which seems to be verified by the anatomical relations of the muscle, has been verified
by the rhinoscopic observations of Dr. Semeleder on the living;
for he first observed a constriction of the mouth of the tube at
each elevation of the velum. Thus the floor of the Eustachian
tube is lifted up at every contraction, and especially that part
of it lying next to the pharyngeal opening; whence it is manifest that a catheter introduced into the tube must, by the elevation of the velum, make a rotary movement. This upward
movement of that part of the floor of the tube next the pharyngeal opening also occurs in connexion with the movement of
swallowing, as the introduced catheter describes the same rotary
movement with the act of swallowing as with the elevation of
the velum."

These experiments can be imitated upon one's own person by
every one, who is accustomed to rhinoscopic self-examinations,
and who possesses some control over the movements of the velum. The other modes of catheterization, such as the introduc-

tion of the instrument through the opposite nostril or through the mouth, &c., have been in part given up, and in part restricted to rare and quite peculiar cases; all these modes of proceeding are, however, frequently so uncertain, that skilful surgeons have thought that they might better be entirely abandoned. The pharyngeal mirror now places the means within our reach, not only of convincing ourselves of the success of the operation, but also of recognising the possible hindrances, and, in difficult cases, of carrying out the operation even under the guidance of the eye.

The author has published in one of the essays cited, a series of experiments bearing upon this point.

I. The catheter of Itard was introduced in the usual manner, carried back to the posterior surface of the soft palate, and the external end handed to an assistant; then the naso-pharyngeal speculum was introduced, and to the assistant was left the carrying out of the various movements, which were necessary to conduct the small head of the catheter into the mouth of the tube.

II. In a second series of experiments, Czermak's rhinoscope was introduced after the catheter had been passed into the naso-pharyngeal space; by this means the examiner had one hand free. After the mouth of the Eustachian tube and the end of the catheter had been brought into the field of view, the proper introduction of the instrument succeeded at once.

For such cases the above described simple method of rhinoscopic examination, without palate-hook, with the mirror and spatula, or with the mirror alone, specially recommends itself. These experiments were made in the presence of Czermak and various other friends, and their favorable result gave great satisfaction to all.

In the earlier attempts there will always be some trouble in consequence of operating according to reflected images. If we look at Fig. 1, Plate I., we see at the left the beak of the catheter in the right sinus of Rosenmüller, and on the right we see it lying in the mouth of the left tube. In both cases the instrument lies actually in a horizontal plane, while in the representation it is projected, and seems to stand almost perpendicularly upon the wall of the pharynx. If the catheter lies in the sinus of Rosenmüller, i. e. behind the mouth of the tube, it seems to stand above the latter, &c. If now we wish the beak of the instrument to glide downwards in the reflected image into the mouth of the tube, the catheter must be drawn outwards, evenly, and horizontally, as if we would withdraw it from the nose, &c. The same is also true of turning. The physician has to occupy himself here at the same time with the illumination, the mirror, and the catheter, and here is again shown the advantage of that method of illumination which is attached to the head of the examiner. Voltolini has constructed a very simple phantom for the preliminary practice of catheterization.

The author contents himself with setting forth the results of the above experiments, and he must leave it with aurists to pass judgment upon the importance to aural practice of this novelty. He can but feel gratified, however, at receiving the favorable judgment of friends who are qualified to judge. Voltolini, an aurist of large practice, pursued the subject still farther in a very pertinent essay, established the fact that even men of distinction are frequently extremely puzzled in the catheterization of the Eustachian tube, giving the causes which may occasion it. He examines first of all the generally acknowledged signs that the operation has succeeded. The following are regarded as such:—1. When the catheter has been passed into the mouth of the Eustachian tube it can be moved still further forward in the direction of its curvature; 2. When the attempt is made to draw the catheter out of the nose horizontally, an obstacle is felt; 3. One sign that the catheter has found the right way, is the feeling of its sliding over the prominence; finally, 4. It has been advanced as an important proof that when the Eustachian tube is open, we can hear air which has been blown into the catheter passing into the cavity of the tympanum.

If we consider these signs a little more closely, they lose much of their apparent exactness. If the sinus of Rosenmüller is comparatively deep, as it sometimes happens, the beak of the instrument may even then be shoved slightly forward. We can feel the presence of an obstacle in an attempt to draw the catheter exactly towards us, when the instrument has not passed into the Eustachian tube, if the sinus of Rosenmüller is narrow and deep, or if the nasal cavity is narrow or constricted, in which case the correct impression on moving the instrument is lost

altogether. Dr. Pollitzer of Vienna has shown the author some
very instructive preparations, in which numerous strong filaments of areolar tissue pass horizontally through the sinus of
Rosenmüller, in consequence of which, instead of a sinus, a
number of little compartments are formed by these bands, into
which the catheter may be directed, and there held fast. It
still remains undecided whether these filaments are congenital
or whether they have resulted from diseased processes. In
other preparations there are numerous enlarged follicles in the
sinus of Rosenmüller, forming deep pockets in which, in like
manner, the beak of the instrument might be caught. The sensation of the instrument's sliding over the posterior prominence
may be rendered uncertain by narrowness of the nasal cavities;
further, the mouth of the tube may lie uncommonly far to one
side, so that the beak of the instrument cannot reach it; finally,
the arrest of development at an infantile stage, as observed by
Voltolini, may occur, in consequence of which the prominences
of the mouth of the tube scarcely exist, and hence are not to be
felt. As regards finally the passage of air through the tube,
Voltolini properly calls attention to the fact that the catheter
is frequently called into use, precisely because the Eustachian
tube is stopped up, and it may happen, if the pharyngeal mouth
of the Eustachian tube lies far outwards, that the beak of the
instrument does not itself enter into it, but remains standing
upon the edge, or stands opposite to it in such a way that after
all, air may be driven in with a forcible current, and thus sometimes, as in the removal of mucus, etc., we may see beneficial
results from catheterization, when, in fact, the operation has not
succeeded. But the circumstances would be different if sounds,
or pieces of catgut, were to be introduced through the catheter:
thus it might happen that the mucous membrane would be injured, and then the air, which is blown in, may not only cause
emphysema of the skin, as happened to Voltolini himself, but it
may produce the most severe accidents, as unfortunately experience can show (Turnbull).

Voltolini gives still further some very good rules for the
performance of catheterization with the aid of the mirror. He
recommends practising the operation with both hands, and in
the catheterization of the right tube introducing the mirror with
the right hand, the catheter with the left, and *vice versa*. The

same rule also applies when other instruments are to be applied in the naso-pharyngeal cavity under the guidance of the mirror.

Dr. Gruber, who has a large aural practice, and Dr. Pollitzer, lecturer upon aural medicine and surgery in Vienna, both in like manner use with advantage the pharyngeal mirror ; and the dissenting judgment of Kramer, who, save in this point, is so deserving of credit, and who seems not to have rightly seized hold of our process, stands yet quite by itself, and need by no means discourage us.

As we can introduce the catheter by the aid of the mirror, so may we equally well apply caustics, and cutting or piercing instruments, and, indeed, we are more frequently placed in the position to make use of them, inasmuch as we can now do so with greater certainty, and can recognise the most exact indications. Voltolini advises the introduction of cauterizing and other instruments through the ear-catheter, when this has received the proper direction towards the given point, and is firmly held by means of Kramer's pincette.

If it is true that three-quarters of all diseases of the ear proceed from the Eustachian tube, or must be treated through this tube, no one will deny the importance of rhinoscopic investigations to aural surgery.

We add a short summary of the pathological conditions hitherto observed at the mouth of the tube. The author himself standing aloof from this branch of science, could collect only a few pertinent observations.

At the mouths of the tubes, in the angle opening towards the nares, formed by the posterior and inferior lips of the prominence (we speak of the image in the mirror) we found frequently deposits of mucus as large as a lentil, without there being any appearance of catarrh, or any closure of the Eustachian tube.

1. Plate I. Fig. 2. In the case of a strong man, 26 years old, who occupied a bed in the hospital in consequence of necrosis of the phalanges, the deposit, above mentioned, was circumscribed on both sides towards the nares, by a dilated vessel of perhaps $\frac{1}{4}'''$ in diameter, running in the reflected image inwards and upwards, in reality forwards and upwards. At first we were almost inclined to take these vessels for the edges of ulcers, and to regard the yellowish crumbling deposit of mucus as an exu-

dation. Moreover some small vessels were found dilated around the mouth of the left tube, and there was also redness of the entire mucous membrane of the naso-pharyngeal space. The figure gives, therefore, a representation of catarrh of the tube, or more properly speaking, of catarrhal inflammation at the mouth of the tube. Only when we questioned the patient, our attention being aroused by this condition, did he acknowledge that for a long time he had heard much more poorly than formerly. The patient escaped further observation.

2. Once we had the opportunity to observe in a young man a chronic catarrh of the naso-pharyngeal cavity, actually at the mouths of the tubes. The mirror showed the surface reddened at those points, the mucous membrane of the prominence of a loose texture like velvet, swollen, and of a dull bluish-red, here and there dotted with small nodules (pharyngitis granulosa), and covering it, in spider lines, were a few small but dilated vessels—especially of venous origin; — at the mouth of the tube itself there was a little thin mucus.

A smaller degree of redness and a larger vessel is shown in Fig. 2, Plate II., at the mouth of the right tube, and this phenomenon may here be explained by the mechanical influence of the existing polyp.

Figs. 5 and 6, Plate I., represent irregular positions of the mouths of the two tubes, but of this hereafter.

We place next a summary of the observations of others upon the alterations of the mouths of the tubes.

3. Czermak. A young man had suffered since his childhood from a discharge from both ears. Afterwards, in consequence of a leap into the water from a spring board, fresh and repeated inflammation of both external ears occurred, and his friends remarked that the patient heard with difficulty. The examination of the membrana tympani only showed a slight opacity. The tonsils and the posterior wall of the pharynx showed in part slight appearances of inflammation, in part traces of previous inflammatory processes. By the application of Valsalva's experiment the air entered readily and freely into the right ear; into the left only after repeated efforts, and in very small quantity. The introduction of the catheter was attended with some difficulty upon the right side; upon the left it passed easily through the nasal cavity, but upon blowing air through it, aus-

cultation discovered a sharp, short, high, whistling murmur, and a crackling and rattling as well upon the membrana tympani as at a little distance from it.

The rhinoscopic observation undertaken without the aid of the hook, showed us the following : The mouth of the right tube was perfectly normal ; on the left side were seen two long tumors, partly smooth, partly shaped like the comb of a cock, proceeding from the lateral wall of the pharynx in such a manner that one lay behind, and the other in front of and below the mouth of the tube ; this itself was concealed, and in like manner the left posterior nasal opening and the ear catheter which had been introduced ; the velum was pressed somewhat towards the right and downwards, hence the uvula stood also to the right, outside of the middle line. Both the tumors were of a darker red than the surrounding parts, and upon being touched with the sound gave the impression of considerable hardness.

Czermak abstains from giving any opinion as to the nature and origin of these tumors.

The author was also enabled a short time since to examine this case, which as yet stands alone of its kind ; he can only confirm in every respect the observation of Czermak, and as for the rest he must also impose upon himself the same reserve.

4. In a case of deafness, which was examined in April, 1860, in the hospital Val de Grace, at Paris, Czermak gained but a negative result, finding merely the vivid redness and slight œdema of the entire pharyngeal mucous membrane.

5. In another patient, who complained of deafness and disagreeable feelings in the upper part of the œsophagus, and who had repeatedly suffered from catarrh of the nasal and pharyngeal cavities, Czermak found deep redness and œdema of the mucous membrane of the naso-pharyngeal space, with abundant secretion of mucus, and tumefaction of the follicles. At about the height of the mouths of the tubes, and concealing these from view, there was a semicircular protruding enlargement of the mucous membrane, which divided the naso-pharyngeal space from behind forwards, into an upper and a lower section as it were. Czermak could not follow the case.

6. In a boy, who was deaf upon the left side, and in whose case the false impression of nasal polypi was given by the digital examination, Czermak found the posterior ends of the turbinated

bones, especially of the right lower one, protruding and much swollen; the mouth of the right tube was normal; on the left side, protruding from the lateral wall of the naso-pharyngeal cavity, there was a mucous tumor almost as thick as one's finger, becoming smaller both upwards and downwards, upon which an irregular indented cicatrix-like depression indicated the extremity of the tube.

7. Observation of Dr. Dauscher of Vienna: A work-woman, thirty years old, had, from a cold, acquired an inflammation of the left middle ear, after the cessation of which deafness remained behind, and was sometimes complicated with subjective impressions of hearing, occurring in paroxysms, to the great annoyance of the patient. The removal of large lumps of wax from the external ear brought no improvement of the hearing. The rhinoscopic examination showed that the mucous membrane at the mouth of the tubes, especially on the left side, was exceedingly reddened. In the mouth of the left tube there was a yellowish grey plug (*Pfropf*) protruding about 2''' from the opening. This condition remained several days without alteration. Itard's catheter was then introduced and an astringent solution injected. At the second injection the patient cried aloud with pain, and complained of great dizziness, and of the impression of a great noise, as if a cannon had been fired off. Soon, however, the patient heard with the left as well as with the right ear. The next day, at the rhinoscopic examination, it was found that the obstruction had entirely disappeared from the mouth of the left tube, while on the lower portion of the same there was a yellowish cream-like deposit to be seen. A complete cure followed the continuation of the astringent injections.

8. Voltolini examined a boy, fourteen years old, deaf in both ears. The result of the external examination was negative. The catheter passed freely into both cavities of the nose, but it was not possible to find the mouth of the tube; only once, and that more by chance, did a current of air pass through the left tube. On the application of the mirror, the naso-pharyngeal space was found to be of an unusual breadth transversely, so that the beak of an ordinary catheter did not reach the mouths of the tubes. The mouth of the right tube was not round, stiff, and open as usual, but there were two lips which lay loose and flaccid upon each other; a small fissure only was left between

them. There was the same condition at the mouth of the left tube, only less pronounced. The mouth of the tube could therefore only be reached with difficulty, and when it was reached, it could not be clearly felt. With the aid of the mirror the operation was performed speedily, easily, and surely, and the result of the treatment was very favorable.

9. Voltolini observed a case quite similar in a girl, eleven years old, and afterwards several similar cases, one of which, a boy sixteen years old, he exhibited to the Medical Society of Breslau. In another case Voltolini found a pharyngeal polyp to be the cause of deafness, but of this bye and bye.

Further Pathological Observations upon the Naso-Pharyngeal Cavity.

The figures 1 and 2 of Plate II. represent some examples of neoplasma, and in particular Fig. 1 relates to a case of

10. Mucous growth, the so-called mucous polyp of the middle and inferior turbinated bones of the left side, in a girl twenty-two years old. Both of the diseased turbinated bones were enlarged, uneven, rough, completely closing up the left posterior nasal opening; the inferior turbinated bone of that side partly overlapped the nasal septum, and the inferior turbinated bone of the right side is thickened at its posterior extremity; some dry greenish mucus clings to the right middle turbinated bone. On all of the turbinated bones a deep blue color was observed, and they were all soft to the touch; the posterior nares and the walls of the naso-pharyngeal space were unaltered.

The examination of the nose anteriorly showed, on the left side, far back, growths of a bluish-red color, uneven, soft, knobbed, and lumpy. The passage of air through the left cavity of the nose was almost completely interrupted. The polypi were removed in the 3d Surgical Ward of the Vienna General Hospital, by the polyp-forceps and scissors; in the course of the operation some small pieces of the left middle and inferior turbinated bones were also necessarily removed; their bony structure proved to be thickened and swollen, and contained considerable fluid. The rhinoscopic condition could not be investigated before the dismissal of the patient, in consequence of external obstacles.

11. Fig. 2, Plate II., gives a representation of a pharyngeal polyp. This was discovered by chance, at one of my class exercises, in a journeyman tailor thirty-five years old, who was in the hospital in consequence of tuberculosis of the testis. The polyp, represented in its real size, covered the greater part of the right posterior nasal opening, and the left almost entirely; it was uneven, of a marked bluish-red color, lustrous, and quite firm to the touch. The point of origin of the structure could not be seen, but everything favored the supposition that the polyp was seated upon the point of union of the hard with the soft palate. The mucous membrane of the naso-pharyngeal space was of a deeper red than its normal condition; a more profuse secretion of mucus at the mouths of both of the tubes was observed; on the right side, between the edge of the nares and the mouth of the tube, there was a small vessel of a line's breadth, half drawn together, and running upwards; both of these phenomena were dependent probably upon the mechanical irritation of the tumor. The naso-pharyngeal space of the patient was in all directions uncommonly large, so that the examination succeeded very readily: otherwise, it would not have been possible to see so much of the nares and of the mouths of the tubes, in spite of the existence of the large neoplasm. The formation did not annoy the patient at all, hence he most obstinately refused any interference with it. Dr. Störk has examined the case frequently, and the author will be well pleased if his drawing is found to be correct.

12. In the course of the summer of 1860 the author saw a young man, nineteen years old, in whom the interrupted passage of the current of air through the nose could be detected at first sight.

The face had a somewhat idiotic expression, from the mouth standing open; the voice was hollow and wanting in resonance. The examination of the nasal cavities anteriorly revealed nothing remarkable. At the first introduction of the nasopharyngeal mirror, which, by the way, was borne without any inconvenience, a body was discovered proceeding from the base of the skull, as large as a hazel-nut, bluish-red, shining, elastic, slightly lobular, which covered the nasal septum, and in great part the posterior nares; but behind, and approximating this, there was still room enough left to see the normal mouths of the tubes and a portion of the base of the skull. The patient's hearing was good.

The case was particularly adapted to the application of the galvano-caustic for the removal of the neoplasm, but it was removed from further observation, as the patient belonged to a Union, which provided for its members in another hospital.

13. Shortly after, the author observed a case precisely similar, in a laundress, forty years old, only the polyp here was lengthened so that its extremity could be seen whenever the velum was lifted up. It was very obvious, in the mirror, that the formation proceeded from the inferior portion of the sphenoid bone. The removal of the neoplasm by an operation was proposed, but the patient did not return.

14. A colleague submitted to a rhinoscopic examination, because he did not inhale sufficient air through the right nostril, and he thought there must be a polyp in the cavity. The external examination showed nothing special. The mirror showed the nasal septum to be turned somewhat towards the left, the right inferior turbinated bone to be much swollen, projecting further backwards than normally into the naso-pharyngeal space; it, as well as the mucous membrane of the septum and of the right nasal cavity, so far as could be seen, was of a deep bluish-red, of the color of an unripe plum, œdematous, hypertrophied.

15. Czermak examined in August, 1860, in La Charité, at Paris, in the presence of the numerous auditory of Velpeau's clinic, a case, in which the existence of a large pharyngeal polyp had been already fully established. The application of his method succeeded, as with most of those persons affected with the larger naso-pharyngeal polypi, with unusual facility, and the lower portion of the polyp could be seen by the clear sunlight quite accurately and clearly in its natural color and condition; on the other hand, the upper portion of the polyp and the seat of its attachment were not visible, in consequence of its large size. The usefulness of rhinoscopy is limited, in cases of such large neoplasms, to ascertaining the superficial character of the accessible portion, and to aiding the eye in the application of the sound, or a sling to the tumor. The patient was first operated upon after Czermak's departure from Paris; otherwise a subsequent examination would have assuredly led to the most

exact conclusions as to the seat of the polyp, the success of the operation, the possible existence of remnants, etc.

16. Voltolini examined a school-boy, seventeen years old, in consequence of marked deafness, especially in the left ear. The external examination gave no satisfactory disclosures. Upon catheterizing the tube, and blowing in air, enormously large masses of tenacious mucus were discharged each time from the opposite nasal opening, even after repeated blowing and clearing out of the nose; still the hearing improved constantly after air was blown in. The disease had existed two years; it had commenced imperceptibly, and had gradually grown worse; ringing in both ears existed at times. The rhinoscopic observation succeeded only upon the second day, and revealed, in the upper part of the pharynx, a polyp, which, itself concealed by a great deal of mucus, covered the mouth of the left tube, and seemed to be of the size of a walnut. In all probability, this was, in consequence of its seat, the cause of the difficulty of hearing, for it on the one hand covered the mouth of the left tube, and on the other, it maintained an irritated condition, and hence also an accumulation of mucus in the region where the Eustachian tubes opened.

Middeldorf removed the polyp by the galvano-caustic. The neoplasm was as large as a filbert. A subsequent rhinoscopic examination showed the naso-pharyngeal space to be clear, and upon the upper edge of the left posterior nares there was a white spot as of a burnt mucous membrane; here manifestly had been the seat of the polyp. Still further there was an evident tumefaction of the mucous membrane of the naso-pharyngeal space. Catheterization was performed several times, and the hearing of the patient greatly improved thereby. Here, also, was the pharyngeal polyp the actual cause of the deafness.

17. Voltolini found in a man seventy-three years old, a large nasal polyp in the left posterior nasal opening, with deafness and a discharge from the ear, probably in consequence of the irritation propagated by pressure upon the tube; in a similar case Voltolini could determine that the nasal polyp had not yet passed beyond the posterior nares.

The two following are cases in which the rhinoscopic examination was made after the removal of naso-pharyngeal polypi.

18. A clerk, 18 years old, a pale weak youth, came under

observation in the spring of 1860, with the statement that he
could inhale no air through his nose. Even four years previous
the increasing nasal character of the patient's speech had excited
attention. Bye and bye, about two years after, the transmission
of air through both nostrils had been rendered difficult. The
patient could only breathe with his mouth open, and the alteration of the voice was remarkable. Swallowing and hearing were
both natural. Abundant hæmorrhage from the nose and mouth
occurred repeatedly. Astringent gargles and injections did not
alter the diseased condition. A year since the polyp was first
recognised as the cause of the phenomena, and upon repeated
applications of the forceps, numerous fragments of the polyp
were withdrawn from the right cavity of the nose; these
together might have been of the size of a small nut. The
repeated and severe hæmorrhages were checked by injections of
a solution of perchloride of iron into the right nostril. After
the operation there was pain in the right ear, and loss of hearing
upon this side; these phenomena, however, passed away again.
Shortly afterwards two small pieces were again withdrawn by
the forceps, accompanied by copious bleeding.

When the patient came to Vienna, the objective phenomena
were as follows: the expression of the countenance, from the
breadth of the ridge of the nose, and from the open mouth, was
peculiarly silly; the respiration was blowing, the tone of the
voice short, weak, and nasal; the velum was broad and high,
tense, but slightly movable, bulging forwards in its entire
extent. Upon introducing the finger, the lower portion of the
septum and the posterior nares were found to be clear, the right
opening much broader than the left, and there was a round,
slightly uneven body, as large as an egg, hanging downwards
from the base of the skull, almost completely filling up the
cavity. The neoplasm could not be seen through the mouth.
The inspection of the left nostril showed it to be normal; that
of the right showed the lower and middle turbinated bones
enlarged in diameter at their anterior extremities, and deeply
reddened.

The rhinoscopic examination gave but very slight satisfaction.
The velum was so stretched and so tense that it could not be
lifted up, much less rolled over. If the uvula was drawn forwards with the palate-hook, the mirror, to be sure, could be

introduced, and the lower circumference of the neoplasm could
be seen, thus establishing also by the mirror the existence of the
polyp; but as this filled up the cavity so completely, it was
impossible to look in towards its base, the only point which was
of importance.

The galvano-caustic was considered as well adapted to this
case, and in the early part of May, 1860, the removal of the
neoplasm was undertaken by Dr. Neumann of Vienna, and
the author. We pass over the details of the operation itself, as
they would carry us too far, and observe only that not a single
drop of blood was lost save a few which escaped in the process
of introducing the wire.

The polyp had a broad oval basis, the longer axis of which
stood almost horizontally; it was about as large as a hen's egg,
and its anterior pointed extremity extended somewhat into the
right nasal opening; from this fact, and from the previous existence of a polyp in the right nostril, the dilatation of the right
nasal opening, and the pressure of the septum towards the left
were explained. The neoplasm was as hard as cartilage, creak-
ing upon being cut, covered with a thin vascular membrane
easily separable; its surface was slightly uneven from numerous
furrows; the face of a section showed in some parts white
ribbon-like streaks, and in others soft masses as large as a pea,
raised above the surface. Sections of numerous vessels generally as large as a poppy-seed, and masses of bone of about the
same size, were visible to the naked eye almost everywhere upon
the cut surface. The microscopic examination showed a tissue
filled with numerous areolar cells, made transparent by the addition of acetic acid, crossing in various directions, and forming
here and there closed spaces.

The course was favorable, and was only interrupted by numerous and considerable hæmorrhages which necessitated the
application of the tampon to the nasal cavities. The patient,
who was very weak and pale after these losses of blood, recovered gradually, and thirty-six days after the operation left
Vienna cured.

The rhinoscopic examination made by the author on the day
of his discharge, gave the appearance represented in Fig. 5,
Plate I. After overcoming some slight difficulties presented by
the tongue, the examination succeeded easily and perfectly;

indeed, one could make a perfect demonstration of the case. The image showed the right nasal opening much broader than the left; the nasal septum shoved far to the left (that is, towards the right in the reflected image); the turbinated bones of the left side pressed out so broadly as scarcely to be recognised; the middle turbinated bone of the dilated right cavity enlarged in its diameter, and covered with bright red fleshy excrescences; the superior meatus, as well as the free space of the nasal cavity, as far as one could see, covered with a dried greenish-yellow pus and mucus; the nasal septum in its lower portion inclined to the right (in the reflection towards the left), but not removed from its inferior attachment. Outwards from the nares on both sides were the images of the pharyngeal mouths of the Eustachian tubes, the left one normal, the right (in the mirror left) elongated, narrowed, and apparently standing higher, but in fact shoved backwards. Above the nares and the nasal septum, stretching somewhat more towards the right, in the reflection towards the left, there was a rough, uneven, puffy surface, having a yellowish-grey color, at the same time moist and glossy, and this was the imperfectly cicatrized base of the polyp, on the lower portion of the body of the sphenoid bone. At the lower third of the septum there was a smooth, white, glistening cicatrix.

The rhinoscopic examination was here easy to carry out in its technicality, but still it was very difficult, inasmuch as at that period one needed ample time to *orient* himself and to be able to determine accurately the individual parts. Let it be remembered that of all the structure of the naso-pharyngeal space, nothing was normal but the mouth of the left tube, and that this therefore could be the only certain point of departure for the examination. Scarcely any one would wish to deny that Czermak's method of examining the naso-pharyngeal cavity is of great value in cases similar to the above. We can be sure that whatever might have been left of the base of the polyp, would at all events be reduced to a minimum by the suppuration following the operation; and also that no other neoplasm existed in the dilated right nasal cavity; had one been found, its removal could have been carefully supervised and conducted by means of the naso-pharyngeal mirror.

With respect to the alteration in form and position of the mouth of the right tube, it should be remembered that the

patient stated that he had suffered previously from deafness in
the right ear for some time, after a polyp as large as a nut had
been torn out of the right nostril; the root of the polyp may
have reached to the mouth of the Eustachian tube, and after its
removal, inflammation may have set in, followed by suppuration,
cicatrization, shrinking, and thus at last by distortion of the right
tube.   At the time of the operation the hearing on both sides
was equally good.   The scar upon the nasal septum may have
proceeded from a slight superficial injury during the operation,
or in applying the tampon.

19. The counterpart of the preceding case is presented by
Fig. 6, Plate I.   The figure presents the rhinoscopic appearance
at the time of the discharge of a lad nineteen years old, after the
extirpation of a naso-pharyngeal polyp, by means of the galvano-
caustic wire.   About a year before, the patient had submitted
to the ligation of a pharyngeal polyp, which had produced an
obstruction of the left nostril, and loss of smell.   Ten days after
the ligature was applied, a lump as large as a walnut was cast
off, and during this period repeated violent hæmorrhages occurred.   Afterwards, still another piece of the polyp was removed
from the left nostril anteriorly, whereupon copious hæmorrhage
again ensued, demanding the application of the tampon.   Immediately before the removal of the neoplasm (May 23, 1860)
the patient presented the following pathological condition.

At the opening of the left nostril lay a soft, elastic, bluish-red
mucous polyp, semi-transparent, and of an oval shape, which
was quite movable and was attached superiorly.   If this was
elevated by a chisel-shaped probe, it was seen to lie upon a second
polyp much paler and more firm, situated upon the floor of the
nasal cavity in such a way, that between the two, the space was
almost entirely filled up.

The finger passed behind the long and flaccid velum, felt quite
a firm body hanging down, about as large as a walnut, rounded
off and having two lobes; it could be completely circumscribed,
and was seated with its broad root upon the anterior half of the
roof of the pharynx and over the nasal septum.   Besides, the
finger felt in front of the polyp, the entire left nasal cavity
filled with an uneven, rather thick polypoid formation (*After-
gebilde*).   The right cavity was somewhat narrowed by the nasal
septum, which was pressed towards the right side; it was other-

wise clear and accessible to the index finger for a considerable distance. The nasal bones were pressed apart somewhat; respiration
through the nose was rendered difficult, especially through the
left nostril; the speech was nasal, the sense of smell almost
entirely absent; the swallowing was not affected.

The examination with the naso-pharyngeal mirror exhibited
the pharyngeal branch of the polyp very plainly as a bluish-red,
shining, round body with numerous enlarged vessels, and with its
surface partially eroded. The mouths of the tubes, the septum,
and the walls of the nares, which a few weeks before were still
partly visible, could no longer be seen; on the contrary, a part of
the pharyngeal portion of the polyp, which at first had been concealed by the everted velum, was now seen.

The patient bore very well the rhinoscopic examination made
with the assistance of the spatula and the hook, for the velum had
become somewhat insensible from the perpetual contact with the
polyp.

The operation was performed at the General Hospital in
Vienna, on the day above mentioned, by Dr. Zsigmondy, the
surgeon in charge of the third surgical ward, and without the loss
of blood, save a slight flow which occurred during the application of the wire. The extraction of the polyp took place in
two movements, inasmuch as the nasal and the pharyngeal
branch must each be removed by itself. We pass over the
description of the operation, as well as many other particulars of
this highly instructive and most accurately described case, inasmuch as they have no special bearing upon our subject. The
entire weight of the portion removed was about four drachms;
the whole mass was about the size of a goose-egg.

After the operation repeated haemorrhages occurred from the
left nostril, so that frequently the tampon became necessary.
Seven weeks after the operation, the patient was discharged cured.
The rhinoscopic examination then made gave the appearance of
Fig. 6, Plate I. The nasal septum was seen pressed so far
towards the right that the right cavity seemed only a crevice
of 2-3″ in width, while the left was much enlarged. At the
base of the skull, towards the left (in the reflection towards the
right), there was a slightly curved bluish-grey cicatrized surface,
the point of attachment of the pharyngeal portion; and upon the
upper and external region of the left nasal cavity, within the

opening of the posterior nares, there was a yellowish-green crust
of dried pus and mucus, the imperfectly cicatrized point of attachment of the nasal portion.   The rest of the mucous membrane of
the naso-pharyngeal space was very uneven ; the mouths of both
tubes were normal ; that of the left was directed somewhat downwards.   That the right nasal cavity should be so much constricted at the time of the patient's dismissal, whilst before the
operation the point of the index finger could be introduced into
it, can be explained by a still further inclination of the septum
from the repeated application of the tampon to the left nasal cavity.

   Zsigmondy says, in his closing observations, the rhinoscopic
results before the operation had undoubtedly in the present case
far less practical value than the examination with the finger.
*   *   *   *   *   On the other hand it is not to be denied, that
the observation of the polypoid growth which brought immediately to our view its color, its superficial vascularity, its excori-
ations, its continued development, and its consequent alterations
in form, gave a desirable completeness to the diagnosis ; that
the rhinoscopic examination incommoded the patient much less
than the manual ; and that it was applicable to a patient with
such special tendencies to hæmorrhage at a time when one
would have hesitated making a thorough manual exploration,
from the possibility of a return of the bleeding.   In like manner the representation brought to our view by the mirror after
the cure, is very valuable, inasmuch as it gives convincing proof
of the completeness of the cure, and of the absence of any after-
growth, thus in fact supplementing the results of the manual
examination.

   20. Czermak observed in Pesth a case which is still unex-
plained.   Dr. P. complained that he had felt for some time an
obstruction in breathing through his nose, and that his voice had
a marked nasal tone.   The rhinoscopic examination succeeded
easily, and showed at once a large body which had been developed, filling up nearly the whole naso-pharyngeal cavity.
Upon closer observation, and upon a tactile examination conducted with the aid of the mirror, this body seemed to be a
vesicle having smooth and thin walls, transparent, quite tense,
and filled with fluid.   Czermak could not examine the case
afterwards, but the practice of self-examination was recommended to the patient.

The figures 3, 4, 5, and 6, Plate II., exhibit cases of ulceration. The marked points of difference of the individual figures
will be readily observed. The author, however, before proceeding to the consideration of the individual cases, would make a
single remark upon ulcerations in the naso-pharyngeal space in
general. He does not hesitate to declare, that as regards form, the
character of the edges, of the base, and of the coating of ulcers
in this cavity, we have no grounds for expressing our views
with any certainty as to their nature and character, at least
according to the present stand-point of the teachings upon ulceration. On the contrary, an exact diagnosis in each individual
case is only made possible by the anamnesis, and by the consideration of its peculiar condition; and even in many cases the
results of a given course of treatment must be first made use of
to confirm the opinion formed, however much uncertainty there
may be in pursuing this path.

21. Fig. 3, Plate II., represents a case of ozœna in a school-boy
sixteen years old. The figure shows an ulceration upon the
upper and inner edge of the left nasal cavity, and upon the turbinated bones of the right side, with loss of substance of the
middle turbinated bone (and necrosis of its bony structure),
and finally a slight infiltration protruding in a point from the
roof of the pharynx. The turbinated bones of the left side are
of a bluish color, and hyperæmic. From the general impression
which the patient made, a scrofulous origin of the disease was
assumed, inasmuch as no other cause for the ulceration could be
discovered, and a corresponding constitutional treatment entered
upon. The local treatment consisted in faithfully cleansing it
with lukewarm water, and in painting it with a solution of one
grain of pure iodine, and one scruple of the iodide of potash to
an ounce of glycerine; this was done both from the nostrils and
from the pharynx; in the latter case by means of a pencil having
a curved handle, which, when the tongue was pressed down,
could be passed upwards behind the velum while it was in a relaxed condition. This treatment persistently followed for several months, had the desired effect. In this patient the palate-
hook could not be used at all, in consequence of the irritability
of the velum; but still the examination and painting were both
very easy when the patient endeavored to utter nasal tones. We
may mention here that Türck has a few times observed circum-

scribed hyperæsthesia of individual parts of the pharynx, which might also give rise to difficulties in examination.

22. A man forty-five years old, of the higher walks in life, was sent to the author to be examined rhinoscopically, as he had been suffering from an obstinate angina. No sooner had the patient commenced narrating the history of his trouble than one could recognise by the tone of his voice that, at all events, some alteration must exist in the nasal cavity. The sufferings of the patient were as follows : For some weeks breathing through the nose had grown gradually more and more laborious, and increasing pain in swallowing, especially dry or fluid substances, had developed itself. Blowing the nose was very difficult, and it had latterly also become painful, bringing to view foul masses of purulent mucus. The respiration was attended with rattling noises in the nose, from the difficulty of cleansing it. The speech had become hoarse, wanting in resonance, and nasal ; and ringing in the ears had been developed.

Upon opening the mouth, the velum was seen constantly elevated and much reddened, the uvula inclining toward the right. An examination of the larynx showed redness of the entrance and of the vocal cords.

The use of the tongue-spatula caused the patient pain in the throat, but it was endured. Scarcely had the velum been somewhat elevated by the palate-hook, and drawn forwards, as the edge of a smooth cream-yellow surface showed itself, with a red and somewhat indistinct edge. The mirror afforded a most striking image, as in Fig. 4, Plate II. The velum was constantly somewhat elevated ; still a space, to be sure rather narrow, was left open between its edge and the posterior wall of the pharynx. The figure shows us, below, the everted portion of the velum, and the upper portion of both nasal cavities, with the superior turbinated bones, and the upper portion of the middle turbinated bones. On the superior and posterior wall of the naso-pharyngeal space, was the large, smooth, cream-yellow surface, towards the right rounded off, expanded out towards the left, and losing itself indefinitely upon the lateral wall. The edges around this surface declined somewhat, were of a bright red color, and were sharply defined upon the vascular mucous membrane of the pharynx. The mouth of the right tube was concealed by the raised velum resting upon the posterior prominence; the mouth of the left tube could not be clearly distinguished in the exudation.

It was clear that we had before us a large ulcer, coated with an abundant deposit. There was no point at which this coating was removed, as the ulcerated surface was so placed in the pharyngeal cavity that it could not by any means be touched by any portion of the structure. As has already been observed, the form of the ulcer reproduced in the drawing can only approximately correspond to the actual.

Of what nature was the ulcer? There could be no suspicion of scrofula, lupus, or tuberculosis, but might not syphilis bear the reproach! The patient had a venereal ulcer twenty-one years before, and in the interval an eruption upon the skin. In each case he had recovered without any special antisyphilitic treatment. The supposition of the syphilitic origin of the pharyngeal ulcer, although highly uncertain, was still the only one possible.

Internal treatment with iodide of potash, a half drachm daily, was ordered, and a rigid diet; locally, painting the parts with the solution mentioned in the preceding case. The next day after the first application of this solution, the appearance of the ulcer was entirely different. The deposit was mostly removed, the surface was covered with little projections, readily bleeding; the edges towards the middle line were sharply cut off. In the earlier days of the treatment there appeared upon the swollen velum small, pointed, red infiltrations, both upon its anterior and posterior surfaces. These became soon more prominent, and of a yellowish color at the point, and acquired a reddened infiltrated halo; afterwards the yellow-colored apex fell off, and there was a small, crater-shaped ulcer, as large as a millet-seed; then the vicinity of this little ulcer was destroyed, and thus by about the fifth day the process had gone on at one point to a perforation of the velum which was large enough to admit a blunt probe. What has been here said of the velum applies equally to the walls of the pharynx, and furnishes us with the history of the origin of many ulcers and the explanation of the small pointed yellow prominences in Figs. 3 and 6, Plate II.

This infiltration of the velum remained stationary, and then disappeared; the perforation, too, became closed, as in a few days the general effect of the iodide of potash was made apparent in the organism. Under the above mentioned treatment, materially aided by an accompanying indisposition, which served to keep the patient in bed, the entire disappearance of all the phenomena ensued within six weeks.

The velum was held constantly elevated as the ulceration extended itself to the right posterior arch of the fauces, and if the swollen velum was dropped down, it must have torn itself from its point of attachment, and a lacerating pain must have been experienced. The seat of the ulcer, and the swelling of the velum, produced the pains in sneezing, blowing the nose, and swallowing, which the patient experienced, descending down along the left arch of the palate, and in consequence of which, from his ignorance of the locality, he considered himself as suffering from sore throat. The elevated position of the velum, firmly maintained with such care, explains also satisfactorily the weak nasal voice.

23. Fig. 6, Plate II., is also without any doubt a well marked case of syphilis of long standing, such as was formerly called secondary syphilis. The representation, taken from a Polish Jew, twenty-three years old, is remarkable from the numerous alterations which have taken place in the structures and which may perhaps make it difficult for many to orient themselves. The specific process had vanished by the use of inunctions, but a reproduction of that which had been destroyed was impossible : hence we limit ourselves simply to the explanation of the representation.

The soft palate is almost entirely wanting; upon the edge of the lost substance at the right (in the figure) there is a bluish smooth cicatrix; upon the left there stand granulations which seem to continue along the floor of the nasal cavity, and from which a fine, reddish, transparent bridge leads to the nasal septum ; the left inferior turbinated bone is wanting, save a stump; the right inferior and middle turbinated bones are partially destroyed ; the mouth of the right tube cannot be well recognised amid the furrows and protuberances which surround it; in the neighborhood of the mouth of the left tube, and upon the external edge of the left cavity, there are respectively small yellow pointed infiltrations; the entire mucous membrane of the naso-pharyngeal space and of the nasal cavities is deeply reddened. Dr. Störk also saw this case frequently. The examination of the nasal cavities anteriorly, showed a perforation immediately at the beginning of the osseous septum.

23. Fig. 5, Plate II., is taken from a young man twenty-two years old, a countryman and companion in faith of the preceding patient. The drawing is made without the application of the palate-hook, and hence we see the uvula, and the greater part of the posterior surface of the soft palate, but yet fore-shortened; the inferior turbinated bones are concealed by the velum; the middle are of a bluish-red. In the vicinity of the mouth of the right tube there is an uneven reddish-colored ulcerated surface, with a slight secretion, by which the entrance to the Eustachian tube is masked; there is a similar ulceration upon the posterior surface of the velum towards the right side (in the figure towards the left); upon the edge of the nasal septum there are three small ulcers, covered with projecting granulations. It was for a long time a matter of difference of opinion whether these three points on the nasal septum were elevated or depressed. The author held the former opinion, and had Czermak upon his side.

It was quite impossible to form an opinion as to the disease which lay at the foundation of these ulcerations. Treatment by inunction had produced no effect, nor had the internal use of the iodide of potash, and the patient returned dissatisfied to his home.

25. In a peasant, sixty-four years old, in whom various signs pointed to the existence of syphilis, but who would not confess to any previously existing disease, we found extensive ulcerations and exudations upon the border, and upon the posterior side of the right palato-pharyngeal arch, and upon both the lateral walls of the naso-pharyngeal space. The ulcerations upon the velum were more easily and more perfectly seen with the mirror than without it. Painting with a solution of corrosive sublimate and afterwards cauterization with nitrate of silver, by means of Leiter's new porte-caustique for cavities, in connexion with the internal use of iodide of potash, gave no satisfactory result. A complete cure only followed a systematic course of inunction.

· The ulcerations did not differ at all from those described in the two preceding cases. Similar observations have been repeatedly made by the author; at one time mercury proved useful, at another iodide of potash, and in some cases no benefit was derived from them. The author makes mention of these circumstances in order to confirm his previous assertion.

Türck publishes three cases of syphilitic ulcerations upon the walls of the naso-pharyngeal cavity, as follows:—

26. An apprentice, twenty years old, came to the hospital three months after he had acquired a genital ulcer. The rhinoscopic examination showed ulcerations upon the roof of the naso-pharyngeal space. Cured by inunction.

27. An apprentice, eighteen years old, had suffered ten days before from a perforation of the soft palate, and a few days before his admission to the hospital a swelling was noticed at the root of the nose. There are ulcers upon the palatal arches, and numerous confluent ulcerations upon the posterior surface of the soft palate, which are far more extensive than those which surround the perforated spots of the anterior surface. There are small mucous growths upon the openings of the posterior nares, and upon the upper wall. Notes of the treatment are wanting.

28. A work-woman, thirty-six years old, had, according to her statement, suffered three years before from an ulcer upon the soft palate, existing without any known cause, which healed under local treatment. For two months past there was again a loss of substance upon the soft palate, and the speech was nasal. Extensive ulcers were found upon the entire posterior wall of the pharynx. Upon the upper wall of the naso-pharyngeal space there was a large loss of substance, and in its vicinity small ulcerations; the mucous membrane, moreover, had everywhere a delicate glandular appearance. In the larynx there was also a large loss of substance, and numerous growths.

The worthy observer draws the following conclusions from these few observations:—

(1.) The loss of substance in the cases of perforation was much larger upon the posterior surface of the soft palate than upon the anterior: the perforation proceeded probably from the former.

(2.) The ulcerations of the naso-pharyngeal cavity were in all three cases not isolated, but accompanied by ulcers of the posterior wall of the pharynx.

(3.) On the other hand, ulcerations were frequently seen in other cases examined by Türck, upon the posterior wall of the pharynx, without any extension into the naso-pharyngeal cavity, or at least, beyond its posterior wall.—From these observations the importance of rhinoscopic examinations in syphilitic patients

is abundantly well proved, particularly as regards the prevention of perforation of the soft palate.

(4.) There are found upon the walls of the naso-pharyngeal
cavity furrows and unevennesses, as well as smaller and larger
round growths, which, perhaps, as the sequelæ of a previous
catarrh, are also found in men otherwise perfectly healthy.
Such phenomena, therefore, occurring as in the last three cases
described, are not to be necessarily referred to syphilis.

Finally, Türck remarks that diseases of the skin in particular
might furnish opportunities for interesting rhinoscopic observations. He also places special emphasis upon the necessity before
examination of cleansing the respective cavities from mucus,
pus, and exudations.

29. Prof. Gerhardt examined with the speculum the nasal
cavities of a patient (who had been for many years syphilitic)
through an elliptical opening in the hard palate 2½ centimètres
(0.98 inches) long, and 1½ centimètres (0.59 inches) at its greatest
width. The smallest laryngeal mirror could be introduced, and
as there was no septum to be observed, it could be placed horizontally with the reflecting surface directed forwards. In the
middle line the reflection in the mirror showed a prong projecting upwards and brightly illuminated, which was so much the
less likely to be recognised as the remnant of the septum, inasmuch as still further forward (appearing, therefore, behind) the
yet uninjured portion of the septum could be seen. Two smooth
round bodies approached this projection from either side; these
were the inferior turbinated bones. Between these and the septum was seen the internal wall of the alæ nasi, which latter, as
rays from the illuminating mirror fell upon the entrance to the
nose, were somewhat more brightly illuminated. The ends of
the hairs at the opening of the nostril were also seen.

30. In a similar case of a syphilitic work-woman, thirty-two
years old, the same observer, examining through a fissure from
3‴–4‴ broad in the velum, found the septum projecting like a
tooth, the nasal meati narrowed, the mucous membrane swollen
and covered with mucus, the right half of the pharynx narrowed
by numerous ulcers as large as a pea, and by horizontal, projecting cicatrices. Gerhardt also presents a third observation of
this kind.

In concluding this division of his treatise, the author would call attention to some departures from the general type, which come to be observed more or less frequently, but which produce no annoyance, and which are only discovered by chance; hence, we cannot describe these as pathological so long as warts and discolored teeth are not generally so regarded. The author would call to mind the form and color of the wall of the nasal septum as represented in the first part of this treatise. When the examination succeeds easily and comparatively well, we find not unfrequently upon the lower third of the nasal septum, near the pale-yellow band, which corresponds to the edge of the vomer, the mucous membrane more or less arched forwards, puffed out, and forming little inequalities. The color of these protuberances is never the deep red of the mucous membrane of the pharynx, which is shown upon the superior third of the septum, but yellowish, shading even to a bluish-grey. These protuberances are of various sizes, their surface sometimes smooth, sometimes finely granular, like a strawberry. These are seen represented, in a very slight degree, in Fig. 2, Plate I. The author found upon himself one such small protuberance, and demonstrated it to others. Prof. Czermak has already observed the same. It is manifest that these "growths" (*Wucherungen*) as the author named them in 1860, for want of a better term, are seated somewhat deeper than the free edge of the septum, even within the nasal cavities. In one case :

31. Fig. 3, Plate I., this "growth" (even in 1862 the author knows no better term) includes the middle and inferior third of the nasal septum, in the shape of two long, lateral tumors of a fine granular surface, situated upon the septum, which, diminishing downwards, unite, and together have the form of the heart of playing-cards. At the same time, the upper portion of the septum is represented as arrow-pointed, smooth, yellow, shining. All the other portions of the naso-pharyngeal space are normal. This peculiar case existed in a young man, twenty-eight years old, who entered the hospital in consequence of an abscess upon the finger.

The author observed a second similar case in an old man. These conditions are so peculiar that the author cannot avoid inviting a more careful observation of similar cases.

32. The author has already hinted in Part I., at the projections of the nasal septum : these may be separated into two

kinds. First, there may be inclinations to one side (and indeed generally towards the left), in which the septum, in its entire height, inclines from the perpendicular dividing plane. These inclinations nowhere show sharp angles, and they contract the space of one-half of the nose in a various degree. They may be congenital or acquired, as in some of the cases described, and only occasionally give rise to functional disturbances, or to hindrances in the introduction of instruments. Such cases have been frequently observed by the author, and can scarcely be mistaken when once they are comprehended.

The other class of deviations from the norm are those in which the septum has one or more projections departing from the perpendicular plane at a more or less acute angle, which, protruding towards one side, are hollow on the other. They stand, according to the cases hitherto recorded, with their greatest diameter in the direction from before backwards, are always congenital formations, and are prejudicial to the development of one or more of the turbinated bones of that half of the nose into which they protrude. The author possesses two preparations of such deviations, but he has not yet been able to prove their existence in the living subject.

Voltolini has compared the skulls of the museum at Breslau, and found that the septum generally inclines towards one side, and indeed most frequently towards the left. The author examined 49 skulls, and found the septum straight 10 times, inclined towards the left 20 times, inclined towards the right 15 times, having an S-shaped curvature 4 times.

33. Fig. 4, Plate I., gives a complete representation of the parts of the naso-pharyngeal space, such as indeed is seldom obtained. It is taken from a patient, who suffered from a fistula of the lachrymal sac of the left side. This case, so interesting in numerous respects, was to have been reported conjointly by the author and his worthy friend, Dr. O. Becker. The latter has as yet been prevented by contingent circumstances from the accomplishment of his part of the task, and the author cannot allow the present opportunity of presenting this remarkable rhinoscopic condition to pass, neither wishing nor intending thereby to forestall Dr. Becker.

The examination succeeded uncommonly easily, without the aid of the palate-hook and the tongue-spatula. This result was much aided by the great distance of the flaccid velum in this patient, from the posterior wall of the pharynx; also it is to be remembered that the floor of the nasal cavity was in this patient more than usually inclined downwards and backwards. The mirror, of 1″ = millimètre diameter,* showed the entire nasal septum, the entire circuit of the posterior nares, the pharyngeal mouths of both Eustachian tubes, as well as the sinuses of Rosenmüller. In the posterior nasal openings were seen on both sides the posterior ends of all three of the turbinated bones in their entire completeness, and further the middle and superior meati, and the passage still above the superior turbinated bone. In the middle meatus of the left side there were a few little warts. The posterior free edge of the nasal septum presented the form of a swelling (fold of membrane) running somewhat obliquely from above downwards, which, especially in the upper part, manifestly projected. This appearance gave us the impression that the vomer was uncommonly short, and the posterior portion of the nasal septum, at least in its lower half, only membranous. In fact this puffed cord could be pushed toward one side or the other by a metallic sound introduced into the nose. This circumstance explained also the great diameter of the naso-pharyngeal space from before backwards, the favorable position of the velum, and especially the facility of the examination.

If a flexible whalebone sound was pushed downwards through the lachrymal fistula and the nasal duct, we could not indeed see its point emerge from the meatus, but when it was still further pushed forward, the sound appeared in the cavity of the pharynx, emerging between the floor of the nasal cavity and the inferior turbinated bone. In the drawing the sound which lay horizontally, and was pushed forward till it touched the posterior wall, is projected almost perpendicularly upon the image of the posterior nares and the upper wall.

The patient died soon after in childbed, at a time when the lachrymal fistula had nearly healed; and we had the opportunity of removing from the cadaver the individual structures of the naso-pharyngeal space. The conjugate diameter at the level of the floor of the nasal cavities, measured from behind forwards to the edge of the septum, was 2.5 centimètres (0.98 inches);

---

* The translator here gives the exact words of the original, but he recognises the discrepancy without attempting to explain it.

to the edge of the vomer 8.7 centimètres (1.64 inches); the
length of the septum, from the anterior nasal openings to the
posterior edge at the same height, was 6 centimètres (2.36
inches); the length of the soft palate, from its point of attachment to the point of the uvula, was 2.5 centimètres (0.98
inches). For the length of 1 centimètre (0.39 inches) from the
posterior edge the nasal septum was formed merely by a membranous fold.

We have frequently gained similar perfect views of the naso-
pharyngeal space, and a similar one Dr. Wagner has obtained
and drawn.

Instruments for Therapeutical Purposes.

For the local application of remedies to the naso-pharyngeal
space, the same instruments which have been devised for the
larynx are used with advantage. Thus the pencil of Dittel,
Gilewsky's tube for blowing powders, the various porte-caustiques
of Leiter, Störk, Zülzer, and others. Finally Störk has invented
a porte-caustique specially for the mouth of the Eustachian tube;
it consists of a vise of platina which receives the caustic, and
which is inclosed and fastened into the tubal end of a catheter
by a screw.

There is however still, in special cases, a wide field left open to
the acuteness of individual observers for new inventions and
alterations.

5

PART II.
LARYNGOSCOPY.

CHAPTER I.

METHOD OF PRACTISING LARYNGOSCOPY.

The Province of the Laryngeal Mirror and a General View of the Larynx.

IF we look into the mouth of a person standing in front of us we can seldom see further along the surface of the tongue than a line which would unite the two anterior papillæ circumvallatæ. Here begins, then, the domain of which the laryngoscope is monarch. We have therefore to consider first the dorsum and the base of the tongue, with the papillæ circumvallatæ and the innumerable irregular follicles. From the base of the tongue the mucous membrane ascends upon the anterior (lingual) surface of the epiglottis, and forms several folds, a middle and two lateral, embracing between them two quite deep pouches; these are respectively the ligamenta glosso-epiglottica and the sinuses of the same name. The attachment of the middle ligament extends more or less upon the anterior surface of the epiglottis along its middle line, and contributes in part to the movability of the epiglottis. This last is more or less pointed, or else cut squarely off, and then often indented in the middle of the free edge, and either flat or rolled in from the sides, forming a gutter opening downwards. From the edge of the epiglottis two thickened folds pass backwards, downwards, and inwards towards each other in a semicircular form, the ligamenta-aryepiglottica, and have peculiar prominences towards their posterior extremities, which are separated from each other by a perpendicular fissure. Posteriorly next the middle line there is found on either side a small, round, pale protuberance which covers the cartilage of Santorini, and about four lines further outwards and forwards there is, on either side, a second similar one covering the cartilage of Wrisberg.

Between these two, which are scarcely ever wanting, especially
in a fully developed larynx, there is found in many persons
still a third less prominent protuberance. From the edges of the
above folds and of the epiglottis the mucous membrane passes
in to the interior surface of the larynx. Here it next covers the
laryngeal surface of the epiglottis, which in the middle is slightly
depressed and has a more or less prominent swelling downwards,
to which Czermak drew attention, and which overhangs the
anterior extremity of the glottis; on both sides of it grooves run
from the edges of the epiglottis downwards and towards each
other. Further, the mucous membrane invests the slightly
depressed inner surfaces of the ary-epiglottic folds, covers the slit
between the two cartilages of Santorini, becomes horizontally
enlarged in passing over the false vocal chord on each side, then
descends into the ventricles of Morgagni, covers then on each
side the vocal chord, and finally passes over the inner surface of
the thyroid and cricoid cartilages, down into the trachea. The
lower edge of the thyroid cartilage as well as the edge of the
cricoid cartilage, and the contour of the tracheal sinus, are seen
as slight projections. At the extremity of the vocal process of
the arytenoid cartilage, where the hyaline structure passes into
the reticular cartilage, and at the anterior extremity of each vocal
chord, also in consequence of the existence of reticular cartilage,
there is a small yellowish fleck (Gerhardt). This corresponds in
the latter case to the cartilage, which according to Bataille, is
frequently found in the anterior quarter of the vocal chord, and
which has already been described by Luschka, being in fact a
thin small cartilage from 2–3 lines in length, which, on either
side arising from the thyroid cartilage, extends into the vocal
chord, consists of reticular cartilage, and sometimes divides itself
like a fork.

On the other hand the mucous membrane of the larynx, as
it passes outwards over the ary-epiglottic folds, invests these
and the posterior surfaces of the arytenoid cartilages, where it
forms numerous horizontal wrinkles, and then passes along into
the œsophagus, the commencement of which is closed and limits
our view. Outwards from the aditus laryngis there is on each
side a large three-sided pouch leading downwards into the œsophagus, its outer walls being formed by the thyroid cartilage
and the greater horn of the os hyoides, and its posterior wall by

the wall of the pharynx. It is called by Betz the *navicular* sinus of the pharynx, and also the fossa laryngo-pharyngea. Besides the above the lower parts of the lateral and posterior walls of the pharynx down to the commencement of the œsophagus fall within the limits of examination.

The examination of the larynx of the lower animals will yield but very little fruit. A careful study of one taken from the cadaver is, however, very beneficial.

Structure of the Larynx.

The framework of the larynx is made up of three unsymmetrical and one symmetrical cartilage, with its delicate appendices. The former are the thyroid and the cricoid cartilages, and that of the epiglottis. Their form has been long well known.

The symmetrical cartilage is the arytenoid cartilage, which has the shape of a triangular pyramid, whose base is placed upon the upper surface of the posterior broader segment of the cricoid cartilage. The inner surfaces of the two arytenoid cartilages, turned towards each other, are both slightly curved from above downwards; on the upper projection, and connected by a joint with it, is a small horn-like cartilage curved inwards, the cartilage of Santorini. In the same manner, from the lateral and external edge of the arytenoid cartilage, there projects outwards and upwards a small staff-like cartilage, the cartilage of Wrisberg. The lower part of the arytenoid cartilage has externally a short process, the muscular process, and anteriorly a longer pointed process, the vocal process. The thyroid cartilage, by its small descending cornua, is so united by a joint with the sides of the cricoid cartilage that a slight movement upon a horizontal axis is possible. The joint between the cricoid and the arytenoid cartilages is capable of very free movement. The arytenoid cartilage, with its hollow articulating surface, directed longitudinally from before backwards, may slide forwards or backwards, outwards or inwards, upon the oval convex articulating surface of the cricoid, and may also turn upon a perpendicular axis.

The individual parts of the larynx are united to each other, and the larynx as a whole to the os hyoides and to the first ring of the trachea by ligamentous formations, which are of various

size and importance.   Besides these, anatomists have designated
as false ligaments of the larynx the three folds of mucous membrane which pass from the root of the tongue on to the anterior
(lingual) surface of the epiglottis, as well as the two which pass
from the lateral edges of the epiglottis backwards, inwards, and
downwards to the top of the arytenoid cartilages, forming the
upper border of the aditus laryngis, and known as the ligamenta
ary-epiglottica.   To this class belongs also the so-called false or
superior vocal chord, a round thick fold of the mucous membrane, which passes from the angle of the thyroid backwards to .
the arytenoid cartilage.   The principal part of this chord is adi-
pose matter and elastic tissue, and, indeed, a broad stripe of the
latter, crescent-like and convex upwards, stretches from the
thyroid to the arytenoid, forming the upper boundary of the
ventricle of Morgagni, and maintaining the form of the superior
vocal chord.   The presence of muscular fibres in this chord is
very doubtful.   Garcia and others assert, however, with injus-
tice, that the superior vocal chords cannot be made to spring
forward to the middle line ; but of this by and by.

Parallel with this chord, beneath it, and separated from it
by the ventricle of Morgagni, lies the true or inferior vocal
chord, which also arising from the angle of the thyroid cartilage, takes a horizontal course backwards, and is inserted upon
the point of the vocal process.   Since it has been ascertained
that the superior vocal chords have no effect in the formation of
the voice, the designation, so frequently made, of "superior or
false vocal" chords has become improper, and for this chord we
shall hereafter in this volume make use of the term *ventricular*
(*Taschen*) *chord*, and shall designate as simply *vocal* chord the
so-called inferior vocal chord, which alone is capable of producing tones.   The vocal chord has upon its free and rather sharp
edge, a stripe of elastic fibres, but consists principally of the
fibres of a muscle of which we shall hereafter treat at
length.

Aside from those muscles which unite the larynx as a whole
with the surrounding parts, we have to consider the actual
laryngeal muscles as follows :—

Three symmetrical, viz. :—

1. The crico-thyroideus, taking its rise from the anterior
segment of the cricoid cartilage, and passing outwards and

upwards to be inserted upon the external and internal surfaces of the lower border of the thyroid cartilage.

2. The crico-arytenoideus lateralis, arising from the upper edge of the cricoid cartilage, posterior to the preceding, passing with parallel fibres, backwards and upwards, to be inserted upon the muscular process of the arytenoid cartilage, and covered for the most part by the plates of the thyroid cartilage.

3. The crico-arytenoideus posticus: arising from the plate of the cricoid, it sends its converging fibres upwards, outwards, and forwards, to be inserted upon the muscular process of the arytenoid cartilage.

Besides these, we have to consider a muscular mass which, having partly a horizontal, and partly, in its superior layers especially, an oblique direction, covers and unites the posterior surfaces of the two arytenoid cartilages, constituting the posterior arytenoid muscle, also called the arytenoideus transversus and obliquus.

The action of these muscles in general terms is as follows: The first draws together the thyroid and cricoid cartilages; it will thus be of importance in making firm the anterior attachment of the vocal chords. The result of the contraction of the second muscle will be a turning of the arytenoid cartilage upon an upright axis, so that the vocal processes of both sides approach each other with their extremities, and at the same time, by the action of the posterior fibres of this muscle, the corresponding arytenoid is turned outwards. As the total effect of the third muscle, we have to regard the turning of the corresponding arytenoid cartilage upon its perpendicular axis, so that the vocal process is directed outwards; this and the former are therefore opponents to each other. The posterior arytenoid muscle will, by the contraction of its fibres, draw the two cartilages together sideways, and it also can roll them outwards.

Thus, following a general usage, in itself bad, we have spoken briefly upon the action of these muscles, which are partly fan-shaped, partly pennate, and which partly, at least according to their position and the arrangement of their bundles, admit of a great variety of motions. Having hinted at this last point, we will return to its consideration in the proper place.

We have still to mention the interior muscle of the larynx,

the thyro-arytenoideus.   According to the able exposition of
M. Bataille, elucidated by numerous drawings, this muscle has
three heads, which have a common origin at the angle of the
thyroid cartilage.   One head (*faisceau plan, Bataille*), trapeziform
in shape, runs with quite horizontal fibres backwards, and is
attached to the lower edge of the vocal process.   It presents a
long, quadrangular, flat bundle of fibres, standing upon its long
edge, and having its surface concave outwards.   The upper edge
of this fasciculus is the sharpest, and is united to the vocal chord
itself (the more firmly the further backwards) by short oblique
fibres.   It forms the lower surface of the vocal chord.   The
third head (faisceau paraboloid, Bataille) sends out its fibres
in different directions; the longest and inferior run backwards and are attached to the outer edge of the arytenoid,
above the attachment of the crico-arytenoideus lateralis.   The
middle and shorter fibres lose themselves in the ligamenta ary-
epiglottica.   The superior, shortest, and most inferior fibres, are
inserted upon the slender thyro-arytenoideus, or pass on to the
epiglottis.   This head forms the chief part of the ligamenta
ary-epiglottica, and the external wall of the ventricle of Morgagni.   The middle or second head (faisceau arciforme—Bataille)
is inserted between the two former heads in such a manner as to
form a triangular surface, whose base rests upon the arytenoid
cartilage, extending, from the upper edge of the flat fasciculus
(the first head) on the vocal process, horizontally outwards to the
lower edge of the paraboloid fasciculus on the posterior external
edge of the arytenoid, while the apex of the triangle lies in
front at the angle of the thyroid.   The surface of this second
head is slightly hollowed from one side to the other, and forms
the lower wall of the ventricle of Morgagni.   Bataille describes a slender laryngeal muscle, which, arising from the posterior external edge of the arytenoid, above the origin of the
muscular process, and passing horizontally forward to the
thyroid, corresponds almost in its direction and position to that
of the ventricular chord.   Others regard this muscle as a continuation of the arytenoideus obliquus.   A large portion of the
middle and posterior fibres of the third head loses itself in this
fasciculus.   Occasionally a fasciculus runs almost in the same
direction from the crico-arytenoideus posticus to the thyroid
and the epiglottis (*Theile's* ary-epiglotticus?); numerous other

variations in the course and connexion of this muscle may also occur. Whether the epiglottis possesses muscular structure in itself, and hence independent motion, is a subject of discussion.

It must still further be remarked that in the interior of the larynx there is a large amount of elastic tissue, *membrana vocalis*, the greater portion of which extends from the anterior angle of the thyroid backwards on both sides, covers the walls of the larynx, and being united in fasciculi, helps to form the vocal and ventricular chords.

The larynx is supplied with blood by two arterial branches. The superior laryngeal artery is a branch of the superior thyroid, is three-fourths of a line in diameter, is accompanied by the superior laryngeal nerve, and enters the larynx through the membrana hyo-thyroidea. The inferior laryngeal artery arises from the inferior thyroid; it is about half a line in diameter, is accompanied by the inferior laryngeal nerve, and generally passes into the interior of the larynx by the membrane between the cricoid and thyroid cartilages.

· The nerves of the larynx on both sides a superior and an inferior, arise together from the N. vagus; the left inferior winds around the arch of the aorta, the right inferior around the subclavian artery, and then at once they pass upwards back to the larynx. The qualities and functions of these nerves still lie in comparative obscurity. The superior as well as the inferior largngeal nerve must contain mixed fibres. Opening and closure of the glottis are both voluntary and involuntary; combined movements are at least very frequent in the larynx; whilst by the exercises of singers and self-observers a peculiar power over certain muscles and parts of the larynx is acquired. It is strange that, apparently at least, the antagonistic muscles in the larynx are supplied from the same nerve-root.

Finally, the larynx and air passages are clothed with a mucous membrane, which, for the most part, has a ciliary epithelium.

Physiology of the Larynx.

The larynx attracts the attention of the physiologist in a three-fold respect. 1st. As a part of the respiratory apparatus;

2d. As the seat of formation of tones and of the voice; 3d. By its relation to the various functions of swallowing, strangling, straining, vomiting, coughing, &c.; *i. e.* its closure.

The Larynx in Relation to Respiration.

In ordinary quiet breathing the glottis is half open, the arytenoid cartilages stand perpendicularly, their apices moderately removed from each other, the epiglottis is depressed backwards so far that it covers almost the entire entrance to the larynx, and only the apices of the arytenoid cartilages and a small fraction of the vocal chords are visible. Between the apex of the epiglottis and the arytenoid cartilages, viewed from the posterior wall of the pharynx, a horizontal elliptical space is left free from the column of air. At each inspiration, when it is not purposely prolonged, there is generally a slight contraction, and then again a dilatation of the glottis. It may also happen, that the apices only of the arytenoid cartilages are directed inwards, in consequence of which the glottis then seems as if bounded by two projecting angles.

According as the respiration is deeper and quicker, or as the necessity for air is greater, will all those parts which stand in the way of the entering air, be involuntarily removed from each other; the epiglottis lifts itself up, the entrance to the larynx becomes wider; the arytenoid cartilages are directed outwards and backwards, their apices stand as far as possible from each other and from the epiglottis, the glottis is wide open and forms an equilateral triangle, whose broad base lies between both the arytenoid cartilages; indeed it even becomes lancet-shaped by the rolling outwards of the arytenoid cartilages, and then its greatest horizontal diameter lies between the points of the two vocal processes. Every change of form in the glottis is the result of quick and sharply defined movements of the arytenoid cartilages.

The Larynx in the Production of the Voice.

The production of the human voice has been long a subject

of investigation by zealous and gifted men. We assert with boldness that the physiology of the formation of the human voice has been greatly elucidated by the laryngoscope, and especially by the labors of Garcia, Czermak, Moura, Bataille, and Merkel. We must limit ourselves here to presenting the actual results which have followed the investigations of these men.

As regards the manner of developing the musical tones of the human voice, and the place of the vocal organ in a given class of musical instruments, the larynx has been described by Johann Müller as a two-lipped membranous pipe (the glottis), with a wind-chest (the trachea), and a mouth-piece, i. e. all those parts which lie above the glottis. The question arises, which is the original producer of the tone, the air or the vocal chords, i. e. whether the tone arises from a regularly recurring interruption of the current of air passing through a tube of a certain calibre, thus throwing the column of air itself into musical vibrations, or whether the expelled current causes the vocal chords to vibrate musically like tongues, these vibrations being imparted to the air. There is good ground for supposing that neither of the two views is exclusively the true, but that both conditions work together to produce the voice. The actual requisites for the production of tone are: approximation of the arytenoid cartilages (a progressive closure of the glottis from backwards forwards), tension of the vocal chords, and finally a current of air of a certain intensity and rapidity. Alterations of one or more of these qualifications disturb the formation of the voice, and produce hoarseness or loss of voice, which are for us the same phenomena, differing only in intensity. On the other hand these different factors have, within certain limits, a compensating capacity, so that a disturbance caused by an altered tension of the vocal chords, by an increase or diminution of the expelled current, may be equalized, &c.

Let us recall how the above mentioned triple factors, necessary to the formation of the voice, are produced. The expelled air (the intensity and rapidity of the current) is a simple respiratory function. The closure of the glottis is produced by the approximation of the arytenoid cartilages from the contraction of the arytenoideus posticus, and of the superior fibres of the crico-arytenoideus posticus, and by their being turned inwards towards each other until their internal surfaces touch, from the

action of the crico-arytenoid lateralis. The tension of the vocal
chords is therefore, 1st. A natural tension from the inherent con-
tractility of living tissue. 2d. A passive longitudinal tension
produced on the one hand by the elevation of the arytenoid
cartilages from the action of the superior fibres of the crico-arytenoideus posticus; on the other hand by the descent of the
anterior attachment of the vocal chords, and indeed of the entire
thyroid cartilage, towards the cricoid cartilage, from the contraction of the fibres of the crico-thyroideus. 3d. An active lateral
tension, from the action of the first and second heads of the vocal
muscle. (Thyreo-arytenoideus—Bataille.) A circumstance of
great interest is the difference between the two registers of the
chest voice and the falsetto. It is well known that these two
registers in passing from one to the other partially lap over each
other, so that a cultivated singer can produce a few tones with
either the chest voice or the falsetto. In such tones the difference in the registers must be the most marked. The above-mentioned requisites are in general alike for the two registers.
Wherein, then, lies the difference? With the chest voice: 1. The
vocal chords vibrate in all their parts. 2. The longitudinal tension of the vocal chords is increased. 3. The glottis is narrow
and straight. 4. The walls of the entrance to the larynx are less
tense. According as the tone becomes higher is the glottis
shorter in its antero-posterior diameter; the approximation of the
arytenoid cartilages is more complete; the view into the larynx
is more free; the epiglottis is elevated. Passing into the falsetto
we may observe that: 1. Only the edges and upper surface of the
vocal chords vibrate. 2. The longitudinal tension is diminished.
3. The form of the glottis becomes elliptical. In deep tones,
especially, the entire larynx and chest vibrate secondarily:
finally, the tone of the voice possesses many shades of difference from the quality of the mouth-piece.

When with a given position and given tension of the vocal
chords the column of air becomes more forcible, the tone will be
higher; we have the means of compensation, however, for these
alterations, and of producing the same tone with various degrees
of force. In a similar manner and under similar circumstances
the entering current may also give rise to tones.

If we consider for one moment speech in itself, we must perceive
that what we generally designate as speech is actually made up of

two parts, of a succession of more or less musical tones, vowels and liquids, and of a series of sounds arising from the closure formed at different points of the mouth, as at the lips, the teeth, the tongue, and the gums. If this closure is suddenly interrupted, or if it exists under such conditions that the current of air produces a rubbing sound, or that the parts forming the obstacle vibrate, then we have the various consonants of different languages.

In this connexion, and especially with reference to the so-called gutturals of the Arabic tongue, Czermak's examination has shown that for these consonants the closure is formed in the larynx itself, a result which has unquestionably greatly enriched our knowledge of the physiology of the formation of the voice.

The whispering tones exist when the vocal chords are only drawn so far towards each other that a rubbing sound is produced, or when, the glottis being closed, the inner surfaces of the arytenoid cartilages stand apart from each other, so that through this portion of the channel the air escapes with a murmur.

Closure of the Larynx.

In swallowing, vomiting, coughing, hawking, and straining, there is always an air-tight closure of the larynx, which may be maintained for a longer or shorter time, or at once removed. When the muscles of the abdomen are in action, the closure of the larynx is only to be regarded as a conjoined movement.

The closure of the larynx occurs according to my experience as follows: 1st. The glottis is closed; 2d. The ventricular chords, by a rapid motion, lay themselves upon each other in the middle line, so that only a small triangular fissure is left posteriorly; 3d. The epiglottis is pressed down upon the entrance to the larynx, inasmuch as on the one hand the base of the tongue is rolled downwards and backwards, and on the other hand the larynx is raised; it is partly effected also by the individual action of the epiglottis.

Czermak has described the closure of the larynx in a similar manner; the complete closure of the ventricular chords is not quite well seen by him, because his epiglottis falls very quickly, and thus its protrusion comes between the chords before they can touch each other. True as is this description of Czermak's, that of Merkel's is equally true, according to whom there is always a space between the ventricular chords. But it is just as true and just as certain that I can close my larynx in a manifestly distinct and successive triple *tempo* in the order above given; that I can produce the two first movements with the epiglottis standing upright, and that I have observed this mode of closing the larynx, and exhibited it an innumerable number of times. To be sure I can also close it rapidly, so that the individual movements, especially the two first, are merged into each other. But I have no ground for supposing that the closure in my case is differently performed whether it occurs slowly or rapidly. It must not be forgotten, moreover, that observers do not generally practise the closing of the larynx in front of the instruments for self-observation. This much results from these experiments: 1. That the ventricular chords are in themselves capable of marked and ready motion, by the action of that fasciculus which Bataille describes as the slender laryngeal muscle. 2. That the closure of the larynx, just as other functions, is attended with slight differences in different people. 3. That we can never establish general laws from the most perfect single observation, and resting upon such observations, no one has the right to accuse other observers of having examined carelessly. To a similar conclusion does the consideration lead, that I produce the Arabic *aïn*, which may be considered as a tremulant tone of the larynx, in a different manner from Czermak, viz. by an elevation of the epiglottis; and yet it is the same sound and of the same value.

A noteworthy observation of Merkel's should be mentioned in this connexion, viz. in gargling, the liquid does not touch the glosso-epiglottic ventricles, the epiglottis, or the posterior wall of the pharynx and larynx, so long as a little of it is not purposely swallowed. Hence we have a proper estimate of the value of gargles in the treatment of diseases of the larynx.*

The Apparatus for Examination.

This, according to the author's method, and for the great

* Von Tröltsch, in his work upon the ear (translated by Dr. Roosa of New York), makes a similar remark, and gives some useful hints upon the proper method of gargling. See pp. 153 *et seq.* of the translation. (Tr.)

majority of cases, is exceedingly simple, and consists of the laryngeal mirrors, the tongue-spatula, and the illuminating spectacles, of which we will speak in the next section.

The mirrors which are now used are almost exclusively of glass, covered with amalgam or with quicksilver; the last are preferable. If from considerations of space, the smallest mirrors become necessary, the author prefers the metallic mirrors (steel), because they are preferable as optical instruments, and nothing of the reflecting surface is lost by even a narrow setting; still, the metallic mirrors are easily injured and difficult to keep in good condition; they rust easily, become scratched in cleaning, are easily stained by the heat, and it is difficult to renew them. The mirrors are attached to a stem of pretty stiff wire at an angle generally of 120°—125° (the angle of attachment), and provided either with permanent handles, or firmly fastened by means of a screw into movable handles. In the quadrangular mirrors the wire is attached to one corner. The author cannot see that any particular form of mirror has any special advantage. The size of the ordinary mirrors is between 1 and 2½ centimètres (0.39—0.98 inches) in diameter. For self-observation and demonstration the mirrors should be from 2½ to 3 centimètres (0.98—1.17 inches) in diameter. The larger mirrors are always to be preferred, as they give a clearer and more perfect image, by means of which the observer finds himself more readily at home.

Tongue-spatulas are used with advantage only under certain conditions. The author makes use of Czermak's modification of the Petit-Simpson spatula, which is applied at a right angle; the lingual leaf is hollow like a spoon, and ribbed upon its lingual surface, and in its long axis considerably bent, so that the free end reaches far down upon the dorsum of the tongue.

The Introduction of the Laryngeal Mirror.

The general rules to which, however, the exceptions are not infrequent, are as follows :—

The uvula rests upon the back of the mirror, and with the soft palate, is pressed backwards and upwards; the lower edge of the mirror is then gently pressed back upon the posterior wall of the pharynx, while its stem lies in the angle of the mouth, behind the superior canine tooth. Quadrangular mirrors should be so introduced that the lower edge rests upon the posterior wall, the upper with the stem against the border of the hard palate, while the two lateral edges should be directed from below upwards. The mirror should stand symmetrically, *i. e.* it should not be turned to one side, but should look exactly downwards and forwards, as otherwise we get an oblique image, and the actual relation of the parts is often first recognised when the hands are changed. Pressure is much more easily borne by the velum and the posterior wall of the pharynx than an uncertain manipulation touching the parts. Every mirror should be warmed before its introduction, simply over the lamp which is used for the examination, to prevent its being covered with moisture. (See note, p. 15.) The warmth of the mirror is also to be proved by touching the operator's own lip or cheek before its application, so that the patient may not be annoyed by a too warm mirror.

Illumination.

[The author has here reproduced the section upon the same topic in Chapter 1st, Part 1st, to which the reader is referred. —(TR.)]

The Laryngoscopic Image.

The image of the larynx in the mirror shows the right vocal chord of the person examined at the left of the observer, and *vice versâ ;* in the same manner all those parts, which in the object stand in front, as the tongue, the epiglottis, the anterior extremity of the vocal chords, the anterior wall of the trachea, are seen in the mirror above and at the same time somewhat forwards; while on the other hand all the parts which in the object lie behind, are seen below and behind in the mirror, as, *e. g.* the arytenoid cartilages, the posterior extremities of the vocal chords, and the commencement of the œsophagus.

The woodcut gives a view of the larynx under very

favorable circumstances; a few words will suffice to make it plain, but we would premise, that so comprehensive a view cannot be obtained by a single position of the mirror, not even if it

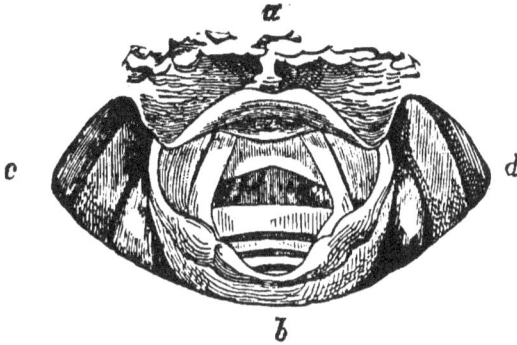

be of the largest size. The parts proceeding in the direction from *a* to *b* are as follows:—The base of the tongue with its papillæ, the region between the tongue and the epiglottis, a portion of the upper surface and the edge of the epiglottis, the protrusion at its attachment, then the widely opened glottis, and through this a view of the anterior wall of the larynx and of the commencement of the trachea; still farther, immediately under the protrusion of the epiglottis, the inner surface of the thyroid cartilage, then the region of the middle crico-thyroid ligament, then a view of the cricoid cartilage, and a few tracheal rings; below, the horizontal mucous folds, which are stretched between the two separated arytenoid cartilages; and finally, the region of the commencement of the œsophagus and the lowest part of the posterior wall of the pharynx.

I have purposely chosen this view, because a representation of the region of the ligamentum conicum has not yet been given, while the anterior and posterior walls of the trachea, and its point of division into the bronchi have often been represented; and because, as it appears, I first recognised the inferior incisure of the thyroid cartilage, and have observed the scar of laryngotomy repeatedly on the living, and in the cadaver have thrown light from the pharynx through the opening for tracheotomy in the above ligament.

The pale mucous fold (commissure) which, when the glottis is open, stretches between the arytenoid cartilages, lies like the folds of a fan and is forced somewhat forward into the interior of the larynx. When the glottis is closed, and the arytenoid cartilages press upon each other, we see between the two arytenoid cartilages only a cleft or furrow, which has been very inappropriately designated by some, the posterior rima glottidis.

In the direction c, d, beginning in the middle, the following parts are seen, viz. the vocal chord forming an angle opening inwards, the apex of which lies at the apex of the vocal process; the entrance to the ventricle of Morgagni, represented in shadow; the ventricular chord, the ary-epiglottic fold, which, shaped like a bow, has an inward and downward direction towards the middle line, and exhibits two projecting nodules towards the interior, namely, the apex of the arytenoid cartilage with the attached corniculi of Santorini, and outwards and forwards (above in the drawing), the apex of the cartilage of Wrisberg. Outwards from the ary-epiglottic fold is the triangular fossa laryngo-pharyngea, or navicularis, which is continuous with the œsophagus, and is bounded as follows: inwards, by the entrance to the larynx; outwards by the thyroid cartilage and the greater cornu of the os hyoides; behind, by the posterior wall of the pharynx; in this hollow we perceive a bright spot, corresponding to a point on the inner surface of the thyroid plate, which is covered with a thinner mucous membrane, so that there the yellowish cartilage shines through; above this there is an enlargement formed by the greater cornu of the os hyoides.

Under unfavorable circumstances the view becomes quite limited, and then one can only see a portion of the base of the tongue, the edge of the epiglottis, the points of the arytenoid cartilages, and a small part of the lower portion of the posterior wall of the pharynx.

Störk describes well the color of the various parts of the larynx under normal conditions. The epiglottis, the interior of the larynx below the glottis, the cricoid cartilage, and the tracheal rings, are generally colored like the conjunctiva of the eyelid. The ary-epiglottic folds, the prominences of the arytenoid cartilages, and the ventricular chords, have about the color of the gums. The vocal chords are of a white, changeable color, and glisten like a tendon. The commissure between the arytenoid cartilages is whitish-yellow or yellowish-grey. The mucous membrane of the trachea between the rings is pale-red ; the point of division and the branches are of a deeper red.   *   *   *   * (A reference to Czermak's photographs closes this section ; it will be found on page 21, Part 1st.—Tr.)

The Examination of the Parts in Detail.

The examination of the base of the tongue scarcely ever presents any serious difficulty. The region between the tongue and the epiglottis, as well as the lingual face of the latter, is frequently very easily and perfectly seen. There is only difficulty in seeing these parts when the tongue, and especially its base, is not pressed forwards ; and in certain cases, when the examination of the epiglottis itself is very easy, where it is forcibly elevated and presses against the tongue. To press down the epiglottis independently upon the entrance to the larynx with the tongue outstretched, and thus to expose the above-named parts, demands, as the author in his own case has experienced, great practice and great control over these organs. Among the various expedients brought forward for the purpose of pressing down the epiglottis, the curved staff of Voltolini affords the best result with the least annoyance to the patient.

The laryngeal surface of the epiglottis is best seen when the view of the larynx, and especially of the anterior extremity of the vocal chords, is the most free : these parts may be observed together when a high note is struck with very little force, or when an inspiration is accompanied with sound upon a high note, or when a sharp and short laugh is given upon a high key, the so-called satanic, scornful laugh. Thus, there is in many cases so favorable a position of the parts that we see the laryngeal surface of the epiglottis more or less foreshortened, with its protrusion ; also the anterior extremity of the vocal chords, and between these and the protrusion of the epiglottis, still a small portion of the anterior wall.

The ready method for gaining a good view into the larynx, consists in making use of the action of the parts themselves.

When a high note is sounded lightly, and even only when the attempt is made without any sound, the glottis is closed and the vocal chords are stretched; the larynx itself is elevated, and is approached somewhat nearer to the observer, and the epiglottis is also elevated as a conjoined act. The tyro will have an opportunity, by the frequent repetition of this procedure of making himself quite familiar with the vocal chords, as they are more easily recognised by their movements; the opportunity is thus offered, too, of observing the closure of the glottis and the tension of the chords, as well as their vibration.

It is well to remember that the note must be sounded full and open, as it is in singing, and not suppressed or shrieking, as a crying child, because then there are coördinate movements which render the view more difficult; it may be that in these latter cases the pharyngeal sac is narrowed in its circumference and the epiglottis is pressed in upon the sides and rolled inwards, a form which Türck likens to a jew's-harp, and which may exist as congenital; the arches of the velum palati may be drawn towards each other; or the dorsum of the tongue is elevated, especially if the note is brought out upon a full vowel sound; or a closure of the glottis follows the note at once, instead of a second inspiration. This happens with those persons who are said to want a natural capacity for singing.

As a rule, the epiglottis raises itself with each inspiration, and the glottis is widened, as has been stated; hence the view into the larynx is often surprisingly free in manifest dyspnœa, if the dyspnœa is not caused by a swelling at the entrance of the glottis. The elevation of the epiglottis is quicker and more marked when the inspiration is quick; it may then be thrown up with a jerk, and the glottis will be closed. The vocal chords also come into sight when the inspiration occurs with sound, and by rapid repetitions of such inspirations we have a succession of tones, which resemble very much the braying of a donkey.

This proceeding sometimes fails even when no pathological causes can be discovered, from the fact that the epiglottis itself fails to rise; but I have only once observed that the epiglottis was, by a quick inspiration, first pressed down upon the entrance of the larynx like a soft cap, and then thrown upwards; this case recalls another in which attacks of suffocation occurred from the epiglottis becoming wedged in the glottis.

The respiratory movements thus induced for securing a better position of the epiglottis often fail entirely, and for various reasons. Many, especially uneducated people, do not comprehend them, even when the same movements are made repeatedly before them; or they become ashamed; some even have refused to put out the tongue because *they thought it was not proper.* Even the first fundamental step of every examination, namely, quiet respiration with the mouth open and the tongue firmly held, must often be first achieved after long practice, during which the observer constantly reminds the patient to breathe, or gives him a musical beat for the inspirations by counting. There is still one other expedient to be considered, which in an unfavorable position of the epiglottis facilitates very much the examination. We may use such mirrors as are more bent towards the staff, like those used in rhinoscopy, in which the angle of attachment of the mirror frequently approaches very nearly a right angle. Thus in an unfavorable position of the epiglottis, the following would be the ordinary and most successful mode of proceeding : Let the patient sigh, that is, let him breathe slowly and deeply, and with the expiration let the voice sound softly and quickly ; also make use of a mirror which is much bent, which can be shoved quite far back and placed in a nearly vertical position. As regards instruments for raising the epiglottis, we would only make use of them very rarely, as in imminent cases or in difficult operations. Such instruments are Voltolini's staff (see above), the forceps of Bruns, with small hooks upon the blades, or the forceps of Fournié which are united to a syringe, and finally Lewin's thimble.

The posterior extremities of the vocal chords, the anterior surfaces of the arytenoid cartilages, as well as the mucous folds between them, and the inner surface of the posterior half of the cricoid cartilage, are all well seen when the mirror is placed more horizontally, and the glottis stands wide open, as the arytenoid cartilages then withdraw from each other, and are directed upwards, backwards, and outwards. The posterior surfaces of the arytenoid cartilages, as far down as the cricoid, are better seen with the glottis closed, as then the apices of the approximated arytenoid cartilages are directed forwards and upwards ; thus the entrance to the larynx is made narrower and the entrance to the laryngo-pharyngeal fossa wider; hence the latter can be seen more easily.

With regard to the examination of the parts situated beneath the glottis, it should be remembered that the light will be the more easily thrown downwards, the more horizontally the laryngeal mirror is held.   But in order to be able to hold the mirror thus, it is well to let the light fall from below upwards upon the laryngeal mirror.   The patient must therefore be so placed that his mouth shall be higher than the eye of the observer, either by placing him upon a stool with a screw, or by elevating his seat.   On this point we agree perfectly with Störk.   Finally, it is only when the local conditions are very favorable, and the axis of the trachea is perfectly straight, that we can look down deeply into the trachea, and can see its point of division even and its right branch.*   These conditions, however, are not so easily fulfilled as one might suppose.   Hence, we maintain that it is always in a measure a matter of chance whether we see the point of division of the trachea, although it may not unfrequently happen.

When the anterior wall of the larynx is shown in the mirror (see figure), and the mirror is then slowly placed more and more horizontally, we can frequently recognise in young persons the inferior thyroid incisure, but always the broad, yellow, projecting cricoid cartilage, and further still, a number of tracheal rings as segments of arches, with their concavity downwards; the narrower they are, the deeper do they lie.   If now the glottis is very broad, or if the mirror is turned towards one side, the lateral wall of the trachea is seen, and then the view is like that of the inside of a long tube in which a screw-thread runs.   Wagner could, after a while, count fourteen rings in his own trachea.   If we proceed from the observation of the anterior wall of the trachea to the point of its division, the rings appear beneath the image of the cricoid cartilage, following each other in their natural order from above downwards, and the point of division first appears below in the image.   The posterior wall of the trachea, although the cartilage rings are wanting in it, shows, according to Türck, also curved lines with an upward concavity; the upper part of the trachea is then seen

---

* In a letter to the translator more than a year since, the author stated that he had twice seen the right bronchus, and once the left.   Gibb mentions two cases in which he saw both; one of these was in a case of paralysis from diphtheria, and this condition he regards as especially favorable for this demonstration.

below in the image, and the point of division appears in the image above and just beneath the epiglottis.

The image of the point of division produces an impression just as if we should see in a mirror a perpendicular projection, clearly illuminated, springing forward, as it were, towards the observer, and terminating upwards (towards the anterior wall) in a bright, triangular surface. On both sides there are dark spots, viz. the commencement of the bronchi. Whether we can also see clearly into the right bronchus or not, will depend partly upon the strength and the proper use of the light.

Obstacles to the Examination, and their Removal.

As we undertake to speak of the difficulties which may be in the way of an examination, we would before all premise that we cast aside altogether every application of force towards overcoming such obstacles. What cannot be attained by calmness, patience, and tenderness, will in no wise be achieved by mere force. The individual to be examined must breathe quietly, deeply, and slowly. This may sometimes only be attained by practice, counting, and persuasion.

When the patient has opened his mouth the tongue should be thrust forwards and flattened, or else hollowed in the middle. Very few persons can place their tongue in this position, and still fewer maintain it when they feel the irritation of a foreign body in the pharynx. As we, in accordance with the maxim given above, reject all forceps for the purpose of holding fast the tongue, we recommend, with Störk, as the simplest, and generally all-sufficient manner, that the tongue should be held at its point by two fingers covered with a napkin, either by the patient himself or by the observer. The tongue should simply be held fast, and neither drawn outwards nor pressed downwards, in order to avoid injuries from the teeth and reflex motions, which readily ensue when the individual to be examined opposes with his tongue the forward movement. In spite of this holding fast of the tongue, its base rises up sometimes like a mountain; in such a case it is sometimes useful and allowable, by means of a fine rod, to press a furrow in the tongue along its middle line. Occasionally it is necessary to lay some charpie or a small compress between the lower incisor-teeth and the frenulum of the tongue. Tongue-spatulas are but seldom applied in the examination of the larynx; for by the use of the ordinary spatula an unfavorable position is artificially given to the tongue, inasmuch as its base is pressed backwards upon the epiglottis, which in turn is pressed down upon the entrance to the larynx, and thus the view of the air-passage is restricted. The above-mentioned spatula, with the well-curved lingual leaf, can, by an elevation of the handle, not only press the base of the tongue downwards, but also forwards, and away from the posterior wall of the pharynx. Rauchfuss showed me a tongue-spatula of horn, which is concave in the middle longitudinally, and hence clings firmly; it irritates but little, and does not reflect light.

The velum claims our attention next. Under favorable circumstances, when the patient opens his mouth, we can see the velum with the uvula hanging freely in the pharynx. When the mirror is introduced (following the middle line of the mouth and pharynx as a guide), the uvula is received upon its back and pushed upwards and backwards. Or the patient may involuntarily draw up the velum and the uvula, and then the introduction of the mirror is much more simple. But sometimes when the uvula is quite long, it happens that its point is concealed behind the tongue, so that on introducing the mirror it does not pass easily under the uvula, but stands in front of the latter; thus we have in the lower part of the mirror an inverted image of the uvula, which conceals the image of the posterior portions of the larynx. We must in such a case either pass the mirror down behind the tongue and draw up the uvula by getting it upon the back of the mirror, preventing by pressure its again falling off, or we may gently rub the uvula with the mirror in order to produce contraction of the velum; or we ask the patient to pronounce "a" (ah, English. Tr.), in consequence of which the velum will be raised; and we make use of that instant to lay the mirror upon the posterior wall of the pharynx, and thus prevent the uvula from falling down again.

When, however, the space between the border of the hard palate and the posterior wall of the pharynx is particularly large, it may happen that an ordinary mirror turns the velum upwards and glides into the nasal cavity, or that when the mirror is applied to the posterior wall of the pharynx, the velum and the uvula fall down in front of it. In such a case the mirror should be only applied to the soft palate, so as to keep the uvula up, or a larger mirror should be used. Such cases as these, if at the same time the soft palate is sensitive, indicate the use of oblong mirrors, which are then to be so introduced that their longest diameter is directed from above downwards and backwards. The same would hold good when the tonsils are enlarged, as an obstacle is thus frequently offered to the introduction of the mirror. If the patient draw a deep breath, the tonsils separate from each other ; we make use of that instant to place the mirror upon the posterior wall of the pharynx. As the tonsils draw nearer together again, they will cover a portion of the mirror, it is true, but another portion of its surface is still left free for use. In this case the advantage, which a narrower mirror possesses, would be lost if it were introduced obliquely.

Among the obstacles presented by the strictures of the pharynx might still be added the rare, indeed, but yet occasional hyperæsthesia of the velum, or of the posterior wall of the pharynx when touched, which render the examination impossible. According to Türck's observation, this over-sensibility is sometimes limited to a very small space. We have found the velum the part most frequently, although after all but seldom, sensitive to the touch. The efforts to overcome this sensibility, by the inhalation or internal use of bromide of potash, have not yielded the desired results, and we are sometimes unable radically to remove this obstacle.*

As to the laryngeal parts, it is only the epiglottis which occasionally presents an obstacle to the examination; it can prevent, or render more difficult, the view into the interior of the larynx. This may arise from its being unusually depressed backwards and downwards, whether it be so naturally or from muscular action, from œdema, from the existence of cicatrices, from a swelling of the ary-epiglottic folds, or simply from the fact that the base of the tongue cannot be pushed forwards. Seldom is this depression so marked that the apices of the arytenoid cartilages cannot at least be seen ; we can then judge from

---

* In the Translator's Appendix to this volume the reader will observe that the author made use of a solution of chloroform and morphia with advantage. (Tr.'

their movability, of the movement of the cartilages themselves, even without seeing the vocal chords.*   Everything relative to giving a more favorable position to the epiglottis has been stated already.   It is of course understood that the glottis must stand wide open if the observer would look down below it.

I have already stated that the examination of the larynx in about twenty-five of every hundred men succeeds at once by the first attempt; that in five per cent. it is altogether impossible; and in the remainder it offers more or less satisfactory results, which, however, may be greatly increased by the patient's perseverance, and his becoming accustomed to the operation.   I have myself made examinations in children only two years of age with satisfactory results, but under very favorable circumstances it is true. With children much patience is, as a rule, needed; and a previous acquaintance is also necessary, so that they may have overcome the first shyness towards the examiner.

Magnifying Apparatus—and Apparatus for Measuring.

See Sections on the same heads in Part I.

Double Mirrors.

Czermak first drew attention to the use of these mirrors, which were united to each other at a variable angle, so as to render visible individual parts which were not accessible to examination with the simple mirror, or only with great difficulty.   Wagner, in New York, has carried out this hint, as regards an examination of the base of the nasal cavity, but the results were unsatisfactory.   Voltolini made use of a double mirror (two laryngeal mirrors united at a right angle, an angular mirror) to get a view of the laryngeal surface of the epiglottis, when the latter was strongly inclined downwards, but the experiment failed because the mirror which was to be brought under the epiglottis must be so very small.   In the same way the double mirror might be applied to the examination of the most anterior part of the glottis. The proposition of Czermak seems theoretically quite feasible;

---

* Gibb states that this depression of the epiglottis exists in *eleven per cent* even among healthy persons.   (Tr.)

but there would always be difficulties arising from the local conditions, and it would always be uncertain, as at one time the observer would have from the simple mirror a single inverted image, at another, from the double mirror, a double inverted image ; and then both images must be compared with each other and with the natural relationship of the parts.

Transmission of Light.

This procedure of Czermak's originally applied to the larynx, and then afterwards also applied to rhinoscopy, is in a few words as follows : A mirror is introduced into the dark pharynx ; and the parts to be examined are illuminated from the exterior by clear concentrated transmitted light. Of course a long neck and delicate tissues are the best adapted to this examination, and those portions only of the air-passages can be seen, which consist not of cartilage but of membranous structure, such as the region of the membrana hyo-thyroidea, the middle crico-thyroid ligament, and lastly the anterior wall of the trachea in so far as it is not covered by the thyroid gland nor by thick layers of the soft parts. The interior of the air-passage is thus illumined with a glowing red light, shading off just as we see it on the edges of the fingers, when they are pressed together and held up between the eye and the sun. However charming and attractive this experiment may be, we are not inclined to recognise any practical value in transmitted light; for if a view into the air-passage is easily obtained, the examination with the illumination from a mirror will certainly render better results; and if the local conditions are unfavorable, the reflection from the larynx illuminated by transmitted light will be just as unsatisfactory as when it is illuminated from above. Still further, the light can scarcely be clear enough to allow one to distinguish accurately existing peculiarities.

Demonstration and Self-Examination.

Auto-laryngoscopic demonstration, a result of self-examination, which has become in Czermak's skilful hands so potent a lever in the elevation of laryngoscopy, is altogether indispensable to profound physiological studies, as has been clearly shown already by numerous passages in this volume. It was the most powerful agent in convincing the most skeptical, even, of the wide reach of the laryngoscope in the physiology, diagnosis, and

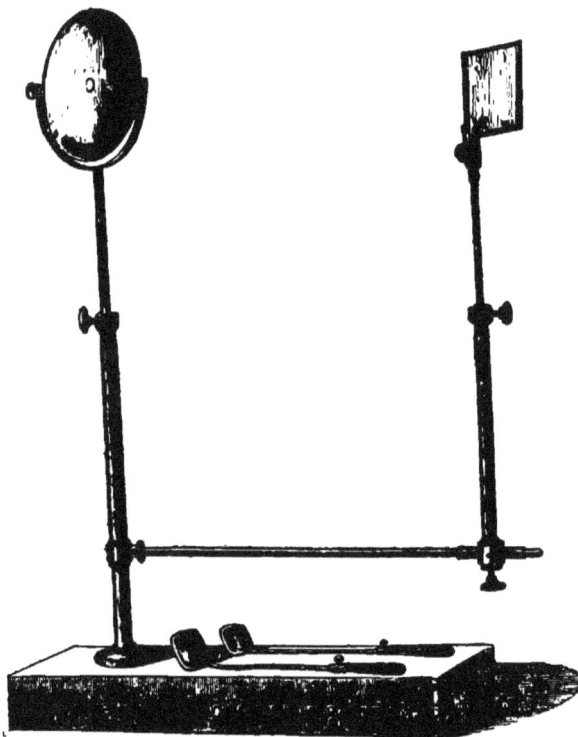

CZERMAK'S APPARATUS.
Taken from his work upon the Laryngoscope.

EXPLANATION.—The round perforated mirror is a concave mirror like the one described by the author. The oblong mirror is a plain mirror, in which the operator himself sees the parts. The round mirror furnishes the illumination, and through the perforation a second observer looks. The light stands at the operator's side a little behind him. The whole apparatus can be placed in the box which serves as a standard.—(TR.)

therapeutics of the larynx. As such, it loses importance in proportion as more and more physicians become sure that laryngoscopy is no deception. The most practised self-observer as such may yet always remain a bungler in the examination and treatment of laryngeal patients. The apparatus best adapted for self-examination and demonstration is Czermak's. The limits of our space forbid us to give in detail its method of application. A common plain mirror of small size is sufficient for self-examina tion. I have always found it very useful in my private course to give, by way of preface, a careful demonstration upon myself.

Demonstration upon another person is a different matter. We are frequently asked, from curiosity or from a desire to verify our statements, to show what we have found to our colleagues. And here it is well to observe the following:

According to what has been already said, the lamp stands at the patient's left, when I am looking with my right eye. The most favorable position for a second observer is at the right and a little behind me, so that he can look close past my illuminat ing mirror. At the same time I must move my head somewhat further from the person to be examined, so that the observer at my side may look in more easily. Finally, I and the second observer can never see precisely the same image, inasmuch as we look upon the surface of the mirror from different angles; when I see the glottis and both vocal chords, the observer at my right sees only the right vocal chord; if he sees the glottis and both vocal chords, I can only see the left vocal chord. De-monstration upon others requires peculiar dexterity.

Examination of the Larynx from below, through the Canula, after the Artificial Opening of the Air Passages.

This method of examination was proposed by Neudörfer in 1858, and was first applied to the living by Czermak in 1859, and thus introduced into practice. To accomplish this, it is necessary to introduce a small mirror with its reflecting surface directed forwards and upwards into a canula with the largest possible window, so that the mirror may serve at the same time for illumination and for observation. The advantage of our method of illumination and manipulation, appears here very manifestly. The mirrors are thin plates of steel, about the size of a lentil, and must, of course, be warmed before introduction. As they cool very readily, and as the examination is thus fre-quently interrupted, Czermak has covered these mirrors with a very thin layer of dissolved caoutchouc, by means of which the image remains clear for a longer time.

The observer sees in the mirror the image of the left vocal chord at his right, and *vice-versâ* (as above), the anterior extremity of the glottis, forwards and downwards, and its posterior extremity backwards and upwards. The lower surface of the vocal chords is not white, but, even under physiological conditions, reddened, as is indeed the whole mucous membrane of the larynx; the vocal chords therefore are only to be surely recognised by their movability; with an open glottis, the lower surface even of the epiglottis may be seen; indeed, in this manner Czermak has thrown light into the pharynx.

This tedious procedure becomes still more difficult from the fact, that we almost always have to deal with pathological altera‧tions, and after the surgical opening of the larynx there may be so much swelling of the mucous membrane below the glottis, that it may, in the form of rolls, narrow the space within the larynx. The worth of the method lies in this, that precisely in pathological cases the epiglottis, either from tumefaction or from the shortening consequent upon the cicatrix, preserves so unfavorable a position, that the examination from above is very unproductive. In this way also, by the aid of the mirror, the sound and other thin instruments might be introduced.

The author has reported a remarkable case of *laryngoscopic self-observation from below through the canula.* A physician from abroad was taken sick here with typhus, which led to perichon‧dritis laryngea; after laryngotomy, and the discharge of a piece of necrosed cartilage, he was so far cured that he could attend to his business, but he was obliged to wear the canula for an indefinite period. He was often examined by the laryngoscopists of Vienna, but a view of the glottis from above was impossible, and even the apices of the arytenoid cartilages were seen very imperfectly and with much difficulty, from the decided and unyielding depression of the epiglottis; nor could the glottis be seen from below, as a fold of œdematous and inflamed mucous membrane closed up the window of the canula; after the repeated removal of small portions, and frequent cauteriza‧tions, it was finally determined to leave this fold to itself; after a while it vanished, and the glottis could now be seen from below in its whole extent, manifestly constricted, but still quite

movable. By an application of Czermak's self-observing apparatus, so that the cone of light should fall above the tracheal
mirror, it was possible for the patient himself to examine the
glottis from below. This patient also gave occasion to numerous
improvements and alterations of the canula, so that they were
adapted to use in speaking.

Under this head belong two observations of Czermak :

1. Balassa performed laryngotomy upon a girl eighteen years old, of a
scrofulous constitution, who had suffered for a long time with continual
pain in her throat and with dyspnœa, which increased sometimes to paroxysms of suffocation. The examination of the larynx from above showed after
a few attempts, that the seat of the constriction was upon the free edge of
the vocal chords, while through the glottis, which gaped somewhat, there
could be seen a shallow longitudinal furrow formed by two dark-colored
protrusions of mucous membrane. The examination from below, showed the
subglottic region of the vocal chords on both sides so swollen with infiltration, that these parts touched inthe middle line. Upon sounding, the
folds seemed very compact. This patient was cauterized from above as well
as from below with the aid of the mirror; dilating instruments were introduced, &c.

2. A girl, eleven years of age, suffering from constitutional syphilis, was
affected with ulcers upon the epiglottis. The operation of laryngotomy
could not be avoided, in spite of the inunction cure, which had produced a
manifest improvement of the laryngoscopic condition. The examination
from the pharynx showed that the epiglottis was swollen almost to the thickness of a finger; on its posterior surface there was a broad ulcer reaching down
deeply; the entrance to the larynx was so constricted, that only a small portion of the swollen vocal chords were visible. By the examination through
the canula, when the glottis was open, there could be seen an ulcerating
surface upon the upper portion of the posterior wall of the larynx, and manifest tumefaction of the posterior part of the left ventricular chord. Moreover,
the apex of the right arytenoid cartilage could be seen, and a portion of the
lower surface of the epiglottis, and through a permanent chink between these
parts one could look up into the pharynx. Improvement.

Œsophagoscopy.

As is well known, the commencement of the œsophagus
lying behind the larynx, is seen in the laryngeal mirror backwards and downwards, beneath the arytenoid cartilages; it is
indeed always closed. The possibility of looking down into the
œsophagus presupposes therefore an artificial dilatation of that
tube.

Attempts of this kind have hitherto been made, so far as I

know, by Lewin in Berlin, and by Voltolini in Breslau. The latter asserts that he can, with reflected sunlight, through a tube a foot long and five lines in diameter, recognise and distinguish perfectly well, objects which lie at the bottom of the tube. In order, then, to admit of an examination of the stomach (!) a tube must be found which can be introduced and then stretched out straight; a requirement which can hardly be fulfilled. Yet with proper instruments it is at least possible to gain a view into the œsophagus a little way down from the top. Lewin has accomplished this upon the living subject, and diagnosed pathological processes with the aid of his œsophageal mirror. I cannot gain access to his original report. Voltolini made experiments for a while upon himself, but gave them up temporarily, as they proved extremely irritating to his throat.

During the past winter I was requested to examine the œsophagus of a man who had suffered great difficulty in swallowing; a bougie which was introduced, met with an obstruction at about the level of the cricoid cartilage. Although after what has just been said, I did not doubt at all the possibility of an examination, yet the whole thing was entirely new to me, and it was necessary to prepare a dilating instrument adapted to the case. I tried first the ordinary pharyngeal forceps with a slight curve; but when the handles stood horizontally, so as to allow a free view of the pharynx, the forceps reached only to the cricoid cartilage. A pharyngeal forceps of greater curvature reached down to the lower edge of the cricoid cartilage, but its application and the opening of the arms so as to dilate the œsophagus, were with difficulty borne by the young man, who had become well accustomed to laryngoscopic examinations. I applied myself therefore to the most willing and accustomed object, and experimented upon *myself.* When I introduced the forceps, I could open them about one centimètre in width at the commencement of the œsophagus, and Dr. Störk was able to see, with the mirror, somewhat further down than ordinarily happened. But as the arms of this forceps were very narrow, the dilatation ensued only from one side to the other, while the larynx pressed itself between the arms, and diminished too much the space from before backwards. Consequently forceps were constructed of similar curvature, but their extremities were made spoon-shaped, like those of lithotomy forceps, so that their diameter at the broadest point, from before backwards, was about 4½ lines or 1 centimètre. Care was also taken to round off thoroughly on all sides the parts which were to be introduced, and to make them smooth, so as to avoid every unnecessary irritation. The handles of the forceps were also bent downwards so as to leave the view into the cavity of the mouth as free as possible. I can bear these forceps myself under certain conditions. I have found that oiling the instrument in no way relieves the unpleasantness of the operation. On the other hand, it is essential to me that the forceps should be well warmed, either by rubbing them between the fingers, by being dipped in warm water, by holding them over a lamp, or even by merely holding the arms for a long time in the mouth.

The forceps need not be introduced so very far, but I must let them glide downwards by their own weight; then they neither strike the posterior wall of the pharynx nor the base of the tongue. To accomplish the examination the tongue must lie in the mouth, well hollowed out. I have gradually acquired the power of pressing the tongue freely down without instruments so far that the point of the epiglottis is visible. Slow, gentle, regular respiration and a firm will are the most effectual means of preventing the movements of swallowing. If these ensue the forceps must be removed, for frequent swallowing with the forceps in the œsophagus is extremely disagreeable. In another case I should have the forceps made weaker, so that they might, if possible, cause even less annoyance. From what has been said, it is evident that œsophagoscopy may indeed prove within certain limits a useful assistant.

With the precautions above stated, the application of the forceps upon the patient referred to, succeeded after a few attempts. I could see as far down as the lowest edge of the cricoid cartilage, i. e. more than an inch down into the œsophagus, and further than that was not necessary in this case. To gain a view still further down it is only necessary to make an alteration in the dilating instrument. Below the points of the forceps a tunnel-shaped cavity is formed upon opening the instrument, and into this we can also look. If a distinct, prominent tumor existed, it would render the examination easier, inasmuch as it would help to dilate the œsophagus. In the present case the result was a negative one, inasmuch as we found nothing in the œsophagus worthy of remark. A carcinomatous constriction sate, as is so frequently the case, in the neighborhood of the entrance to the stomach; and the obstacle to the introduction of the bougie at the level of the larynx, arose from irregular spasmodic contractions of the muscles of the pharynx, a circumstance not chargeable to the examination. As a rule, our patients will be much better contented when we do not find "*interesting cases.*"

CHAPTER II.

I by no means propose to give a systematic view of diseases of the larynx from a laryngoscopic stand-point, and indeed at present it is hardly possible to do so; even in making this beginning I become conscious for the first time of numerous difficulties.

Their divisions and classification must experience many alterations as the result of manifold observations, which can be made alone upon the living; new types of disease, I might almost say, have been discovered, now that we are in a position to perceive alterations, no trace of which can be found upon the cadaver, or in which only the very last stages of development there meet us. The pathology of the larynx, we must say, has gained a rich prize from the use of the mirror; and this has partly been made possible only by the fact, that we could correct our previous ideas upon many of the physiological relations of the air-passages and of the parts surrounding its commencement. The therapeutics of the diseases of the larynx, is, so to speak, newly created; on the one hand we have gained new and established indications; on the other hand local treatment has passed out of the condition of deceptive groping into that of manipulation which can be counted upon and watched over.

The amount of material at hand is sufficiently large and varied, yet it is incomplete and much scattered. While some forms of disease have been thoroughly elaborated, others have passed almost without a single word. If I undertake the task, it is from the desire partly to place in review and in order the abundance of material lying before us, and partly to add, as I hope, from my own observation and experience, something new and valuable. I must consider, by the way, certain affec-

tions of the pharynx which sometimes give occasion to examinations with the mirror.

The Diseases of the Mucous Membrane of the Larynx—Hyper-æmia, Hæmorrhage, Anæmia.

As to these conditions in the mucous membrane of the air-
passages of the living, no clear light was to be obtained before
the laryngeal mirror was brought into use as a means of examination.   The more acute forms of simple hyperæmia, in particular, are seldom presented to anatomical observation.   It is not
always very easy to distinguish between simple hyperæmia and
certain forms of catarrhal inflammation, if the secretion in the
latter is very slight.   Hyperæmia occurs either from slight irritation, or as a concomitant of other diseased conditions, and it
occasionally leads to a hæmorrhage of slight extent either into
the tissue or upon the free surface.   The blood is either mingled
in the form of streaks or small clots with the secretion which is
hawked or coughed up, or it is diffused in the secretion, impart-
ing to it a local brownish color.   It is stated that hyperæmia of
the mucous membrane of the larynx occurs in connection with
anomalous menstruation, with suppressed hæmorrhoidal dis-
charges, with abnormal conditions of the heart, with scorbutus,
tuberculosis, etc.; these are most generally the chronic forms,
while the more acute forms arise as a rule from local irritants.
The hyperæmia of catarrh and of ulceration will be treated of
hereafter.   Hyperæmia sometimes affects individual parts, sometimes it is general.

The cases which I have observed are as follows:

When self-observation or demonstration upon myself is continued longer on an average than a half hour, I perceive, in
addition to redness of the velum, a slight redness of the entrance
to the larynx, extending down beyond the ventricular chords,
joined to a slight feeling of heat, dryness, and scratching, which
soon vanishes after swallowing a little cold water and after a
short rest.   The same is true in the examination of others; the
time requisite for the appearance of such alterations is various;
sometimes it is only first seen after more than an hour's examination.   Upon the vocal chords this hyperæmia appears, though

seldom, as a rosy shade, upon the other structures as an increase of their reddish tinge. The exciting causes here are manifestly the drying of the mucous membrane by the more abundant and more rapid changes of the air, the effort in producing certain movements and positions; and, more important still, the quality of air breathed, and the condition of the laryngeal mucous membrane at the time; for when there is smoke in the air which I am breathing, the phenomena appear much earlier, and so when I have just smoked. The appearance of this hyperæmia is an indication to break off the examination.

In the same manner, a hyperæmia of the mucous membrane of the larynx, affecting the vocal chords themselves, has been observed both by me and others, in persons who have been singing for some time. In a case where the pharynx and the œsophagus were burnt by a solution of caustic potash, which I saw shortly after the injury, I found, in addition to eschars upon the posterior surface of the arytenoid cartilage, and upon the apex of the epiglottis, hyperæmia also of the entire epiglottis, and of the anterior surface of the arytenoid cartilages, as well as of the ventricular chords. Gerhardt mentions that in his own case, even after taking a warm cup of strong tea, he has often suddenly felt as if a foreign body were in the larynx, accompanied by a desire to cough and a hoarseness lasting several hours; all of which phenomena vanish by the following morning. Laryngoscopy revealed each time merely a swelling of that fold which, by the opening of the glottis, stretches between the arytenoid cartilages, and by its closure lies in radiated folds.

A general well-pronounced redness, and in particular a rosy, evenly developed tinge of the vocal chords occurred in the course of a protracted laryngoscopic examination, in the case of a patient who was suffering from measles, and who had at the same time a slight laryngeal catarrh, by which, however, the vocal chords were never colored. (STOFFELLA.)

A strong vigorous man in the prime of life was seized, without any known cause, with severe cramps in the stomach and nausea after a hearty dinner. After repeated attacks of violent and exhausting vomiting the patient became hoarse, and noticed bloody streaks in the sputa. The laryngoscopic examination which was made the next day, showed a deep flocculent redness on the vocal chords of both sides, especially of their attachments and of the ventricles of Morgagni. The ventricular bands, as well as the remaining parts of the larynx, differed very little from their normal color. On the anterior quarter of the right ventricular chord, facing the cavity of the larynx, I found a clot of blood of quite a bright red color and as large as a hemp-seed, which, in spite of coughing and hawking, maintained its position. I suppose that the hæmorrhage had occurred at this spot, and that the hyperæmia and bursting of the vessels was to be ascribed to the altered pressure upon the parts which attended the intense action of the abdominal muscles.

A short time since a man, about forty-five years old, came to me complaining of tickling and a slight scratching in the larynx, which compelled him to clear his throat frequently. There was no cough, the voice was slightly altered, no real pain. The sputa was raised without difficulty, and consisted of lumps of mucus with some air-vesicles. Sometimes there were streaks of blood mingled with the sputa, but in the morning its color was a brownish-red. For his own peace of mind he wished an examination. I found in the pharynx varicose veins and a few granulations. The epiglottis, the entrance to the larynx, and the ventricular chords had quite a normal appearance ; the vocal chords, a little way from both their anterior and posterior attachments, were slightly reddened. Upon the anterior extremity of the left ventricular chord there was a small hæmorrhagic spot, and upon it a slight fibrinous clot. I considered this point as the source of the hæmorrhage, and explained the appearance of the red streaks in the sputa during the day by the pouring out of fresh blood when the throat was cleared, while the blood which oozed out in the course of the night would give the reddish color to the morning discharge in which it was diffused. These appearances lasted for three weeks. No particular cause could be discovered. The patient had about that time suffered severe mental emotion, and stated that he had been subject to hæmorrhoids, but had lost no blood for three weeks, and that it often happened that for months none would come away. I refrain from drawing a conclusion upon these data.

Gerhardt found the laryngeal mucous membrane cyanotic, with a diffuse bluish shimmer in emphysematous persons.

Some years ago I examined a man who complained of disagreeable feelings deep down in the pharynx, or as people generally say, in his throat. I found the mucous membrane of the velum injected and red, and so also that of the posterior and lateral walls of the pharynx ; the mucous membrane of the fossa laryngo-pharyngea participated in this to such a degree that the light points with which we have become acquainted could not be discovered ; moreover, I found in the right fossa from twelve to fifteen hæmorrhagic spots, of the size of a millet-seed, standing at slight distances from each other.

The disagreeable feeling in hyperæmia of the laryngeal mucous membrane is not easily explained ; and the alteration in the voice, which is scarcely ever wanting when the vocal chords are involved, but which is often difficult to establish, as so few men sufficiently observe their own voices, can only be explained in this case by an alteration of the tension and capacity of vibration of the vocal chords depending upon the increased fulness of the vessels themselves. It is evident that under this head much yet remains to be elaborated.

Anæmia of the Larynx,

which is only to be distinguished in its more developed forms, may occur under all those circumstances which generally give rise to it. I have observed it very often and well marked among tuberculous patients, where no other disease of the larynx yet existed, and also where it was associated here and there with hyperæmia and ulceration of other parts. Störk makes the same observation.

Catarrhal Inflammation of the Air-Passages.— Czermak, Gerhardt, Lewin, Störk, Türck.

This may be acute or chronic, primary or secondary, with the general well known character of catarrh of the mucous membranes, viz. redness, tumefaction, and increased and altered secretion. While we hold in reserve the consideration of secondary symptomatic catarrh, we proceed to treat here of the *primary acute catarrh of the air-passages.* It is of course understood, that in many cases not only the mucous membrane but also the tissue lying beneath it, suffers ; hence the disease of the sub-mucous tissue must also be considered.

The redness is either equally distributed, or a network of variable thickness is formed in the diseased mucous membrane by a greater or less number of injected vessels; moreover all grades of redness come to light, from pale rosy red to the deepest red of the buccal mucous membrane, and even to a bluish-red discoloration. The tumefaction depends partly upon the greater fulness of the vessels, partly upon the infiltration of the tissue with the exudation. Thus the mucous membrane, which in normal conditions is smooth and but slightly lustrous, receives a watery lustre, reflects the light strongly, or else it has a velvet-like appearance with a dull lustre, owing to the great sponginess of the tissue. The secretion is in most cases increased, but a difference is to be observed in this respect according to the seat of the inflammation. The more excessive secretion of a serous fluid, as we frequently find it upon the mucous membrane of the nose and eyes, I have hardly been able to detect in the larynx. Only when the secretion has become more consistent and muco-purulent, is it found collected at certain points in greater or less quantities. Hæmorrhage may doubtless occur with catarrh, but I have never observed it. Störk cites such cases. The causes which may lead to catarrhal inflammation of the air-passages have been long known.

The individual portions of the larynx can be diseased by themselves alone, or in various combinations, and in various degrees; thus the epiglottis, the ary-epiglottic folds, the coverings of the arytenoid cartilages, the ventricular and vocal chords, may be either separately affected, or all at the same time. It is well known that the mucous membrane of the pharynx, of the nose, of the trachea and its branches—are all frequently diseased in connexion with the larynx, either simultaneously or continuously, inasmuch as a causal relation exists in respect to the extension from one of these parts to another. According as the catarrhal inflammation affects individual or rather certain parts, are there special phenomena to be observed.

The alterations of the epiglottis comprise all stages between a mere injected redness, and its transformation into an immovable stump of the thickness of a finger (epiglottitis). The redness and swelling of the epiglottis affect both surfaces, or merely one, and more frequently the laryngeal surface. I think I have already remarked that this is frequently dependent upon the initial location of the disease; for I have frequently found in laryngeal catarrh that only the under surface of the epiglottis was reddened, when the mucous membrane of the pharynx was not diseased, and that on the other hand in catarrh of the pharynx the lingual surface of the epiglottis actually participates. Störk makes the same observation. The disease, especially the tumefaction, also affects frequently but one half of the epiglottis. The inflamed epiglottis is very sensitive to immediate touch, and in like manner pressure from without over the os hyoides, produces pain. The epiglottis when greatly swollen presents an obstacle to the examination of the interior of the larynx.

The catarrhal inflammation of the mucous membrane covering the arytenoid cartilages is manifested by redness, swelling, and frequently also by an altered motility; in some cases the swelling is so great that the cartilages seem like round, smooth knobs as large as a hazle-nut. Pressure on both sides in the region of the cartilages, thus moving them towards each other, causes pain; so also the pressure of the larynx against the vertebral column. The posterior wall of the section of the larynx placed above the glottis is often at the same time diseased.

The relations of the ary-epiglottic folds and of the ventricular chords are similar; they may be diseased independently, or in connexion with the epiglottis and the mucous membrane of the arytenoid cartilages; on one side or on both, according as the ventricular bands are swollen, will the ventricles of Morgagni be diminished, the vocal chords concealed, and their motility affected.

The vocal chords are in part or wholly injected or reddened. The coloring exhibits all shades, from the palest rose-red to the deepest darkest red; not unfrequently there are found isolated dark flecks (ecchymoses). Sometimes marked tumefaction can be observed. More rarely, manifest long red streaks are seen, either on the free edge or on the ventricular surface. The partial redness proceeds generally either from the attachments of the vocal chords or from the ventricles. The vocal chords sometimes seem thickened by tumefaction, especially in such a manner that the edges are not as sharp as in the normal condition. Türck says that in some cases he has observed an appearance on the vocal chords and upon other limited parts, as if they had been lightly touched with a dilute solution of nitrate of silver. This probably arises from a spongy structure, and from a partial loss of epithelium. The point which corresponds to the vocal process is indeed the most rarely reddened; frequently this yellow spot is seen very clearly in consequence of the redness of the vicinity, and it might very easily be mistaken for an ulcer, especially when the secretion remains attached there.

In the trachea the redness and tumefaction are seldom so marked that the cartilages do not shine through it.

The catarrhal secretion lies especially at certain points, particularly between parts which rub against each other, or in cavities;—thus in clumps, on the under surface of the epiglottis, and upon the arytenoid cartilages, as well as upon the anterior attachment of the vocal chords; also, in very small clumps upon the edges of the vocal chords. If the secretion is viscous and stringy, threads of it are sometimes found stretching between the epiglottis and the arytenoid cartilages; between the two arytenoids; also in a horizontal direction between the vocal chords, especially in their anterior part; and between the points of the vocal processes. In the trachea, the secretion is often accumulated in great amounts at certain points.

The functional disturbances which occur in connexion with catarrh, are cough, pain, difficulty of swallowing, alteration of the voice, dyspnœa, sensibility to external pressure. Although coughing in itself is no characteristic phenomenon of laryngeal catarrh, yet its relation to the disease of individual parts of the larynx is by no means sufficiently investigated. The pain experienced in coughing, and also without the latter, is readily explained by the inflammatory tension, as well as by the movement upon each other, of parts tumefied and partially deprived of their epithelium; the latter point explains also the irritation which dust, cold air, and cold or acrid drinks produce. Pain in swallowing occurs, because thereby a closure of the larynx is produced, by means of which certain parts are pressed closely upon each other, as e. g. the points of the vocal processes, the arytenoid cartilages, or the epiglottis, and the cartilages of Santorini. It may also happen that, in consequence of the swelling of individual parts, no sufficient closure of the larynx is effected, and so, whatever is swallowed, comes in contact with parts which are not touched in a physiological condition, and upon which therefore it acts as a foreign body ; as e. g. in tumefaction of the epiglottis, or of the mucous folds between the arytenoids, by which a fissure remains between the two. Lewin observes, that small ulcers upon the ary-epiglottic and glosso-epiglottic ligaments produce sometimes difficulty of swallowing, and sometimes obstinate coughing, which are wont to yield at once to local treatment. The dyspnœa is explained either immediately by the swelling of the vocal and ventricular chords of the ary-epiglottic folds, or of the mucous membrane upon the posterior wall, or by the difficulty of raising the secretion ; or mediately, inasmuch as by the swelling of the ventricular chords, or of the mucous membrane around the arytenoids, the necessary dilatation of the glottis is prevented. The alterations of the voice, hoarseness, or aphonia, are scarcely ever wanting, and open a fruitful field of research ; in consequence of their great importance, and of their presence in almost all laryngeal disease, these phenomena deserve special consideration.

Its Course.

In very slight cases the catarrhal appearances amount only to injection, and these cases are well again in a few days. When there is an even redness, especially upon the vocal chords, the disease continues at least two or three weeks. In the higher grades of inflammation, even after a few days, ulcers are formed. Such ulcers are generally superficial, and will heal when the inflammation ceases in the surrounding parts without any farther treatment. Ulcers of this kind have thus far been observed upon the epiglottis, near its free border, where it presses upon the points of the cartilages of Santorini in swallowing; upon the arytenoid cartilages, and those of Santorini; upon the points of the vocal processes; and finally upon the anterior angle of the glottis,* and upon the vocal chords; generally in the latter case, as might be supposed, upon both sides, and of almost equal development.

Türck has observed ulcers in fully developed catarrh, once upon the covering of the arytenoid and Santorini cartilages, where they were so small that he needed a lens to distinguish them ; once upon the posterior section of the left vocal chord ;

---

* The papillary growths which are frequently found at this point are significant of . these ulcers.

and once in greater extent upon the anterior wall of the larynx. Their course was short, and kept pace with that of the catarrhal inflammation.

Störk once found a large flat superficial ulceration upon the inner surface of the left arytenoid cartilage, above the attachment of the vocal chord, between the two arytenoids, and immediately below the posterior attachment of the vocal chord. There was a greyish-yellow deposit upon those spots where loss of substance had occurred. The diagnosis of the ulcer, occurring in connexion with catarrhal symptoms, wavered between syphilis and catarrh. Störk decided upon the latter, from the slight loss of substance after so long an existence, and his opinion was verified by a speedy cure following very mild treatment.

I have had one case of this kind which I will proceed to record : A colleague of mine, a man of about twenty-six years of age, and of a tuberculous tendency, with dulness in the apices of the lungs, began to feel pain upon speaking and swallowing without knowing any particular cause for it ; he thought he was attacked with laryngeal phthisis. The laryngoscopic examination showed the epiglottis, especially in its lower surface, intensely and evenly reddened ; so also the mucous membrane of the arytenoids. Upon the lower surface of the epiglottis, about three lines from its edge, and almost in the middle line, there was a small nodule of the size of a lentil, oblong, projecting, and yellow, which in swallowing came to lie exactly upon the summits of the arytenoids, and hence the pain on swallowing. Somewhat to the left and below this, there was a second small nodule as large as a hemp-seed, and of the same color as the surrounding mucous membrane. The pain upon speaking was produced by the vibration of the swollen epiglottis caused by the current of air. The next day the first nodule burst and brought into view a very flat red spot, with loss of substance and with flattened edges ; the second one had become yellowish, and also burst in the course of the second day. Both healed conjointly with the catarrh without any treatment, leaving no trace behind, and at the same time the pain vanished. The most important point for the patient was my assurance that there was no tuberculosis.

Türck describes as simple ulcers many which are first observed after the accession of the first symptoms of disease, and with which the symptoms of catarrh are wanting. Hoarseness and coughing existed; pain was almost entirely absent; a cure rapidly followed an expectant treatment.

The acute catarrh is cured, or passes into

Chronic Idiopathic Catarrh of the Larynx.

This is either developed from the acute form under improper management, or from the continued operation of external irritants or constitutional conditions; or it may have developed itself at once without being preceded by any acute symptoms. Türck here draws attention to partial inflammations of the vocal chords, which pursue a tedious course without pain, and also without a tendency to cough, but sometimes with a very slight alteration of the voice. My opinion is that much of this form of catarrh will after a while be recognised as symptomatic.

The color of the mucous membrane in chronic catarrh is sometimes a bluish-red; it is often scarcely altered during the more free intervals, or the alteration is limited to a few spots; the mucous membrane is covered not unfrequently with granulations of the size of a millet-seed, and these are generally the most obstinate cases; in some there is a superficial ulceration; the secretion varies greatly both as to degree and character. There is a frequent recurrence of acute catarrh upon the slightest causes. The alteration of the voice is recurring frequently or it is continuous, while in acute catarrh it lasts but comparatively a short time. Pain and the desire to cough are almost entirely wanting. Difficulties in swallowing occur in granular laryngitis; dyspnœa is found in the more severe cases. The farther alterations are either new growths of areolar tissue, continued swelling of individual parts, hypertrophy, or more rarely atrophy of the mucous membrane as well as of the sub-mucous tissue of the muscles, and of the cartilages even, inasmuch as chronic catarrhs generally extend in depth. Thus there may be a considerable shrinking of the cartilage in consequence of a catarrh (particularly noticeable upon the epiglottis), just as there is in the cartilages of the eyelids in trachoma. Not unfrequently papillary growths, swelling of the follicles, and the formation of small mucous polypi, succeed catarrh.

Hypertrophy of the vocal chords is shown essentially by

thickening, especially on the edges, which, when the glottis is closed, no longer spring forward so sharply as before; and besides, the color is more or less reddish, at least yellowish. Atrophy of the vocal chords is often seen in their thinness merely, more frequently, in the formation of longitudinal folds, which are especially visible on attempting to make a sound ; the color of the vocal chords becomes yellowish; the edges particularly are thin and flabby.

Chronic Symptomatic Catarrh of the Larynx.

This is developed as a secondary disease in the vicinity of syphilitic and tuberculous ulcerations in the larynx, or in connexion with cancer, and it may produce a callous-like hardening of the mucous membrane and of the sub-mucous tissue (Türck). I think that a closer investigation of this form of disease will afford much that is new. Secondary catarrhs are certainly more frequent than is generally supposed. It will therefore only be a question, in many cases, of finding out the correct relationship. Thus I pronounce a large number of the cases of chronic inflammation of the vocal chords symptomatic; and for the establishment of my assertion I refer to the following cases, selected from a large number of observations.

A young lady of seventeen years of age, who had begun to train herself as a singer, observed for some time that she became hoarse after very slight exertion of the vocal organs, and this gave occasion to a laryngoscopic examination. The lady had previously resided in the country, and rejoiced in the robust health of country life. For the last year and a half she has lived in Vienna, and has for several months complained of numerous troubles, as loss of appetite, excessive and too frequent menstruation, alternating with a slight mucous discharge. The figure is quite well developed ; the face is pale, blushing easily ; and the muscular development weak.

The laryngoscope showed anæmia of the epiglottis, which was, however, streaked with a few injected vessels, remarkable pallor of the entrance to the larynx, of the ventricular chords, and of the interior of the larynx. The vocal chords, especially in front and towards the ventricles of Morgagni, were reddened, and their mucous covering slightly œdematous. Upon the point of the vocal processes on each side, there was a large, and upon the anterior extremity of each vocal chord a small, oblong, yellow fleck, corresponding to the described cartilaginous point, over which the firmly adherent mucous layer was not injected; upon the edge and the upper surface of the vocal chords, there was mucus of a glassy appearance, in small clumps; the closure of the glottis was rather slow, and the longitudinal tension seemed to be diminished in uttering tones. There were a few paroxysms of coughing and an extremely slight discharge in the morning; and after more severe exercises of the organ, there were unpleasant feelings not definitely marked. The patient had been treated by various physicians for her general condition.

The disease of the larynx might be a simple catarrh, or it might have arisen from over exertion. I did not discern at once its true character. I considered as the actual cause of the disturbance of the voice, a weakness of the vocal muscles, which I attributed to her general unhealthy condition. I therefore sent her for the summer to Vöslau, recommended moderate exercise in the air, and gave internally the sesquichloride of iron dissolved with the succinate of ammonia and syrup of orange-peel, gradually rising to high doses. I think I have observed that of all the preparations of iron, this one is the best tolerated, and that the succinate of ammonia is an admirable adjuvant. Locally, pulverized alum was blown into the larynx twice a week, but without special effect. In the course of treatment it appeared that at the time of the catamenia, each period being accompanied by a decided increase in the secretion from the larynx, the injection of the vocal chords increased to a red flesh color, in consequence of which the yellow flecks seemed still more dazzling. In the same manner the injection was much more marked if the patient had been singing for a half hour before the examination, after which time the voice was evidently poorer.

In the course of the summer her general health became very much improved; the menses observed the regular period, lasted four days, and the discharge was bright red; the mucous discharge had ceased; the whole figure became more full; a slight injection of the vocal chords continued and was increased by singing. The movement of the arytenoid cartilages was sharper; the tension of the vocal chords had become normal.

8

For the last three weeks I have ordered from 30 to 40 inhalations of a pulverized solution of tannin daily (in parts of 1 : 100), and I observe a general improvement; in particular, I find after each sitting the injected redness of the vocal chords diminished, and the secretion coagulated in small white flakes on the edges of the vocal chords, but especially between their anterior extremities and adhering to the points of the vocal processes, thus fully explaining the hoarseness which vanished upon the inhalation.

I am satisfied with the results of the treatment thus far, and lately I have increased the inhalations to three times thirty daily. I think after what has been said that my view of the relation of the laryngeal phenomena to the constitutional trouble must be admitted. Recently the young lady has been able to resume her exercises in singing to her great satisfaction. Other similar cases I have observed, but not so carefully. I shall speak of symptomatic catarrh in connexion with ulceration in due time.

Croup and Diphtheritis.

There are treatises enough upon these processes from a pathological, an anatomical, and a therapeutical point of view. Laryngoscopy has thus far very little to offer, and hence we treat of them both under one head ; and indeed the distinctions from a laryngoscopic stand-point are not very marked. Observations have been made upon croup and primary diphtheritis. Türck says he has never examined any patients with croup. Störk speaks casually of croup, but mentions nothing farther upon the subject. Diphtheritis occurs secondly in exanthematic processes, as typhus and tuberculosis. Türck describes a case of the latter sort. Circumscribed diphtheritis (aphthæ) are found in the larynx under like circumstances, and have been frequently observed.

Störk describes a case which must be regarded as circumscribed croupous inflammation :

A young lady twenty-four years old, suffered for some time from an irritation and tendency to cough, and from hoarseness, without any known cause. The organs of the chest were normal. The laryngeal mirror showed a thick exudative deposit beneath the vocal chords, and also upon the interior of the larynx, which so far diminished the laryngeal space in circumference as to cause frequent attacks of suffocation. Spasmodic coughing ensued, by means of which the exudation was sometimes expelled. A ring-shaped mass of greyish-yellow color, compact and viscid, was formed at some points, hard like an eschar, and streaked with blood; it consisted of mucous globules, pus cells, and fat, without manifest structure. According to the inspection, and from the breadth of this exudation, the disease must have extended beyond the cricoid cartilage. In other respects the larynx was perfectly normal. A cure followed the cauterization with a concentrated solution of nitrate of silver, although frequent relapses occurred.

Fauvel publishes a case of Moura's: A man had recovered from varicella, but with difficulty of deglutition and hoarseness, and with swelling of the glands of the lower jaw. There was œdema of the entrance to the larynx, with constriction. The epiglottis was bluish-red, and there was an ulcer on its anterior surface on the left side; otherwise it was for the most part covered with quite a thick yellowish-white pseudo-membrane. The left arytenoid was swollen as large as a hazel-nut, dark red, and covered in front and above with a thick yellowish-green false membrane with flat and indistinct edges.

The author has, by the kindness of a colleague, had the opportunity of examining a case of diphtheritis in the general hospital. The patient was a girl twelve years old, greatly wasted away at the time of the examination, and suffering greatly from difficulty of breathing. The examination of the pharynx showed that the mucous membrane, upon the lower half of its posterior wall as well as upon the palate and its arches, was in a marked degree bluish-red and hyperæmic; in some places it was destroyed, and in others covered with small ecchymoses and a few greyish-yellow shreds. The examination with the laryngoscope was quite easy; it revealed a condition similar to that just described upon the epiglottis and upon the entrance to the glottis; the vocal chords were colored yellow; but below those, and quite deep down, as far as could be seen, the mucous membrane of the trachea was covered with a greyish-yellow, dirty, dingy deposit contracting the space; here and there it hung in shreds from the sides; it was about one line in thickness, and through it the surface beneath could be seen, having a deep bluish-red color and a spotted uneven appearance. The diagnosis of diphtheritis was easily made, and it appeared that the process had commenced above and had extended itself downwards. The patient died at the end of two days, and the post-mortem examination revealed diphtheritic exudations in the trachea as far down as the division of the bronchi ; at some points the exudation had been discharged. I think no physician can err in the laryngoscopic diagnosis of diphtheritis and croup.

Türck describes a case of diphtheritic ulcers upon the arytenoid cartilages and upon the posterior wall of the trachea in connexion with tuberculosis, but the case was not examined with the mirror.

Aphthous ulcers, in connexion with advanced tuberculosis, have been frequently seen by nearly every one of us. I recall some cases in which they sate upon the arytenoid cartilages, upon the posterior wall of the larynx, and once upon the anterior wall of the trachea at about the height of the sixth cartilaginous ring. Diphtheritic ulcers are also found in the larynx in connexion with typhus. They are also observed upon the soft palate and upon the posterior wall of the pharynx in very advanced cases of tuberculosis.

Diseases of the Sub-Mucous Connective Tissue.

I grant that in most cases the sub-mucous connective tissue is only diseased secondarily, either from the mucous membrane or from the cartilage ; but on the one hand cases occur in which diseases of the sub-mucous tissue exist after the most transitory symptoms elsewhere; on the other hand, the diseases of other tissues often first assume a particular significance, upon involving the sub-mucous tissue ; and hence I find myself obliged to present the following section.

The Inflammation of the Sub-Mucous Connective Tissue.

This is, in most cases, the concomitant of other processes, most frequently of catarrh, but it does occur, although rarely, in connexion with temporarily diseased conditions of the mucous membrane and of other tissues, and must hence be considered as an independent disease. In particular under this head, I might class the chronic infiltration and swelling of the entrance to the larynx, of the vocal chords, of the interior of the larynx, and of the trachea, either in connexion with chronic catarrh (Türck, Czermak), with syphilis (Czermak, Gerhardt, Türck, Störk, the author), with scrophulosis (Czermak, one case), or with Lupus (Türck, Gerhardt, the author). The inflammation of the sub-mucous tissue may lead, however, to a circumscribed or diffused formation of pus, as well as to hypertrophy. Hence, we arrive at the consideration of

Laryngeal Abscess.

This disease, although upon the whole rare, is in most cases dependent upon perichondritis. Abscesses are also developed in consequence of inflammation of the glands, or of the mucous membrane; or they are situated upon the exterior of the larynx. Only a few abscesses of the larynx are therefore to be classed under this head. Such cases have been mentioned frequently. Lewin opened with a knife constructed by himself, an abscess upon the protrusion of the epiglottis, which had caused aphonia.

Serous Infiltration.

This is seen in the larynx under the same circumstances as in other parts of the body, viz. as a concomitant of inflammatory and ulcerative processes; thus of catarrh, especially where there are pharyngeal affections (Rühle), seldom of croup, and especially seldom in children; farther still, it is seen in cases where there is an obstacle to the recurrent venous blood, as in diseases of the heart, and of the large vessels in aneurisms; also in cases where there is a general watery condition of the blood, as in the stages of convalescence from all severe diseases. Still farther, œdema of this nature is seen in connexion with diseased processes in the vicinity of the larynx as a collateral condition; or with inflammation of the areolar tissues of the neck, or of the parotid gland, just as œdema of the eyelids may occur when a furuncle has been developed in the temporal region; or with perichondritis, just as in other parts of the body we find œdema of the overlying tissues in connexion with deeply seated formations of pus and with periostitis. But, as has been stated, there are forms of œdema of the sub-mucous tissue in the larynx, in which diseases of the other tissues are almost or entirely wanting; and these forms will next be considered as simple œdema.

Œdema affects particularly parts having a spongy structure, as the ary-epiglottic ligaments, and the ventricular chords; more rarely the epiglottis or the coverings of the arytenoids alone. Œdema in the larynx is sometimes on both sides, sometimes on one, and is more common in adults than in children. The simple appellation œdema of the glottis must be done away with, as the vocal chords are, on the contrary, but very little inclined to become œdematous; all the cases of simple primary œdema occurred after " taking cold."

I recall two lads, school boys, between ten and twelve years of age, whom I had an opportunity to examine in the spring of 1859 at the "*filial*" hospital in the *Leopold-stadt*, in Dr. Lewinsky's department. It was in March or April, and the air was very sharp and rough. After taking cold, both became sick, having a slight cough, no expectoration; a rough non-resonant voice, slight sensitiveness of the larynx to the touch, and disagreeable sensations in speaking and swallowing; when breathing quietly there was no annoyance felt. The examination with the mirror showed in both cases that the ventricular chords on both sides were œdematous, tense, of a bluish pale red color, shining, slightly transparent, and protruding as swellings of the thickness of one's little finger, so far into the cavity of the larynx that the edges of the vocal chords were barely visible on uttering sounds. There were no farther manifest alterations in the larynx. The stiffness and the brilliancy of these swellings caused the appearance of great tension. In one of these cases it was possible to reach these swellings with the point of the finger, and feel of them; they were thus found to be quite tense; after the introduction of the finger the patient felt more comfortable, perhaps because the swelling was partially removed by the pressure. If it had been necessary, I would have scarified these tumors with the aid of the mirror; a cure followed simple treatment, although very slowly.

Störk describes similar cases of œdema of the epiglottis, of the ary-epiglottic folds, of the ventricular chords, and of the arytenoid cartilages, also arising after colds, but accompanied by catarrh of the mucous membrane; but to this malady alone, so high a degree of œdema could hardly be attributed. A similar case is also related by Czermak.

Numerous observations have been recorded of œdema accompanying convalescence, ulceration, tuberculosis, and typhus, as well as typhous perichondritis. In a case of the last kind, Türck found a high degree of œdema of both vocal chords. I do not think that we can recognise the character of the œdema with the laryngoscope. Cases of infiltration of blood (Rühle) do not seem as yet to have been observed.

Treatment of Inflammation of the Larynx.

It would not indeed be in place here to repeat all that has been everywhere said and that may be read upon the prevention and treatment of laryngeal catarrh, croup, &c. From our standpoint we must consider more in detail the local treatment of the affections of the larynx, as this is the only reliable method, and has only become possible and certain since the introduction of the mirror.

Czermak, making the eye the guide to the hand, first used the sound with the aid of the mirror in the larynx, and applied local treatment; he devised this method. Störk, too, has rendered service in the matter of cauterization. The remedies hitherto applied locally with the aid of the mirror are almost entirely caustics and astringents. They are applied in substance, in solution, and in the form of powder, corresponding to the various modes of application. Numerous peculiar instruments have been brought forward, upon which we shall have something to say. The application of solid substances had not once been attempted before the introduction of the mirror. Störk describes this process very well.

Hitherto laryngoscopists have used ordinary nitrate of silver and that which has been softened by the addition of potash; caustic potash; Vienna paste in sticks; crystallized chromic acid; the dry perchloride of iron; and the ioduret of the chloride of mercury (Boutigny's salt). The action of the caustics is manifold, viz. astringent, changing the character of the secretions, exciting reflex action, and causing local destruction. If it is necessary merely to cover a raw spot, and to protect it in the act of swallowing, the modified nitrate of silver answers every purpose. Where there is a greater loss of substance,

which must be filled up with granulations, that preparation is too
weak, and the pure nitrate of silver, of such inestimable value,
comes in play.   For the destruction of neoplasms and infiltrations, the lunar caustic should be allowed to work mechanically
also; but this does not succeed well in the larynx.   In such
cases caustic potash or Vienna paste are better, although
from the collateral hyperæmia induced, they cause great pain
and dysphagia, which last several hours; but the destruction is
accomplished more rapidly, and hence time is gained.   The
Vienna paste in stick has the advantage of liquefying less than
the caustic potash.   The salt of Boutigny liquefies slowly, cau-
terizes very deeply, and produces severe pain.   In cauterizing
the pharynx, and the entrance to the larynx, the superfluous
caustic should be at once neutralized—nitrate of silver with a
solution of salt, and caustic potash with oil.   For the purposes
of destruction Lewin applies especially nitrate of silver combined with chromic acid.   The latter destroys very rapidly.   It
must be kept in well closed vessels, as it dissolves very easily.
The small crystals are applied by means of a properly curved
pincette.   The application of this remedy demands the greatest
caution, as, if it reaches the stomach in the smallest doses, it
produces gastro-enteritis.   Lewin therefore prefers not to enter
very deeply with this remedy, but he considers it especially useful in hypertrophy of the tonsils.   For the alteration of the
tissues Lewin employs solutions of corrosive sublimate, and of
the sulphates of copper and zinc.

It is advisable not to use very strong solutions at first, and
never to cauterize without having had previous diligent practice
upon the cadaver or the phantom.   These cauterizations are
accomplished with the aid of the mirror, which is introduced so
as to illuminate the part to be cauterized.   Then the *porte-caustique* is carried up to the mirror, and the cauterizing substance is
turned towards the spot to be treated, and then the first favorable moment is seized to carry it home.   From the instant that
the spot has been touched, nothing more can be seen as a rule,
and both instruments must be withdrawn quickly and adroitly.
We must proceed in the same way if upon entering we touched
the wrong spot, or if during the process a movement of choking
or swallowing occurs, which the operator does not follow with
sufficient readiness; in these motions the larynx ascends and

closes itself, the tongue passes backwards, or is in choking pushed forwards. So the motion of swallowing may prevent our drawing out our instrument; the pharynx slides up over the caustic, and a subsequent examination always shows us that we have touched the region between the tongue and the epiglottis. This region and the base of the tongue are the most easily reached, and will often be touched involuntarily, an accident which may still happen after much practice. Another difficulty consists in becoming well acquainted with individual parts in the image of the mirror. The reaction after the cauterization is often for a moment terrifying; it would be well to forewarn the patient of this in a measure.

Gerhardt is of the opinion that the paroxysmal coughing and choking may be useful, inasmuch as in subparalytic conditions they excite powerful contractions, or in partial œdema, they aid its removal by pressure. The secretion of the parts touched is at first increased; afterwards there ensues a feeling of dryness and heat. The secretion of the mucous membrane of the mouth may also be considerably increased. Gerhardt observed once sugillations and a bloody expectoration after cauterizing a condyloma; once after repeated cauterizations of the epiglottis, he saw a circumscribed croupous exudation which lasted six days, and once numerous diffuse grey spots.

The epiglottis is easily cauterized; on its lingual surface, by simply touching the epiglottis as dexterously as one hits a nail with the hammer, and by pressing it down (horizontal cauterization, Störk); and upon its laryngeal surface by carrying the porte-caustique down to the mirror, allowing the instrument to sink forwards and downwards, and with it pressing the epiglottis against the tongue (perpendicular cauterization). Coughing, tickling, difficulty in swallowing, and if the cauterized surface is large, suffocative paroxysms also ensue. In such a case give the patient cold water to swallow. The ary-epiglottic folds are cauterized perpendicularly by pressing the instrument towards one side or the other, and by sliding it forwards or backwards according to necessity. The cauterization of the arytenoid cartilages is horizontal. Pain upon swallowing and tickling follow, but no suffocative paroxysms. The ventricular and vocal chords are cauterized perpendicularly like the surfaces of the interior of the larynx. The consequences are spasm of the

glottis, terrible anguish, and dyspnœa; the patient thinks he
will suffocate, throws himself about, and seizes hold involuntarily of any fixed object in order to inspire more freely.   These
phenomena continue one or two minutes, and pass away upon
drinking cold water.   Coughing soon follows, and it is sometimes so violent that capillary hæmorrhages occur—hence,
caution must be observed in individuals where hæmoptysis is to
be feared.   If the cautery does not adhere, it is often partially
thrown into the physician's face by coughing.   Some individuals
who breathe in very exact time, and who suffer from various
forms of ulceration, bear the cauterization even of the vocal
chords without any special reaction.

The cauterization of the interior of the larynx, below the
vocal chords of the thyroid and cricoid cartilages, is to be undertaken cautiously, as considerable swelling readily follows.   It is
difficult, as we have to pass through the glottis while it is standing wide open, without touching it.   On withdrawing the instrument, the vocal chords may clasp the porte-caustique.   The
sequelæ are pain, a scraped sensation, and short, suffocative
paroxysms, in consequence of touching the vocal chords in the
withdrawal.   Where the pain continues, cold applications should
be made to the larynx.   If we wish to cauterize anteriorly, the
instrument must be drawn forwards, and the handle raised; thus
the beak sinks forwards and downwards.   If we wish to touch
the posterior wall, or the space between the arytenoids, the
instrument must be directed backwards and the handle is
depressed.   For cauterizing the sides of the larynx, the instrument may be shoved along on the surface, or allowed to glide
from above downwards.   I do not believe in cauterizing the
trachea with the solid caustic.

In order rightly to understand these directions, taken for the
most part from Störk, it is necessary to be acquainted with his
porte-caustique.

This instrument consists of a metallic catheter tube, which has upon the
beak a movable cap provided with a large fenestrum; the other end is
attached to a short wooden handle; above, the tube has a slip, running in
which is the slide of a spring, resting upon a long spiral screw.   At the
curvature the screw is attached to a chain, which reaches into the movable
cap, and carries the porte-caustique, a metallic spoon in which the cauterizing substance is placed.   By a movement of the slide the cautery can be

turned upon its axis. On introducing it, the fenestrum is turned towards the desired spot, the cautery being turned in the opposite direction, and when the proper point is reached by a movement of the slide, the cautery is brought to the fenestrum of the cap, and pressed upon the spot. Lewin complains that the point of the caustic pencil cannot be used with this instrument.

The laryngeal parts endure but a momentary pressure; hence the caustic should protrude freely forwards in order to accomplish a sufficiently energetic cauterization by a momentary pressure.

The first porte-caustiques which Czermak made use of were simple clamps on a bent wire. Störk has squeezed the caustic into the blades of ordinary curved dressing-forceps by means of gutta-percha, and thus applied them.

I consider the following the requisites of a good porte-caustique, and Rauchfuss agrees with me:—It must be possible to introduce the cautery covered; it should be freely applicable in all directions by a slight sliding movement, and should stick securely so that none of it can fall down. Leiter has sought to fulfil by his porte-caustique these demands as well as still another, viz. that the operator should be able to give the instrument any desirable curve so as to adapt it to various points.

The lapis is stuck in a little round cupola-like basket of platinum wire; a round silver cap is made to cover or uncover this basket by the movement of a slide in the handle, which slide works upon an easily moving combination of balls strung upon a flexible wire attached to the platinum cup. But the lapis very easily becomes loose in its cup, and the instrument loses strength by its flexibility. This also, like Störk's instrument, is thick, and takes up much space and light.

Zülzer had a porte-caustique made with the ordinary catheter-curve, consisting of a thin metallic tube which becomes somewhat wider at the beak. In this tube a wire runs, on the end of which a button is placed; upon this button the nitrate of silver is fastened by dipping the button into a mass kept at a melting point in a little cup over the flame of a spirit-lamp. The button of melted caustic is covered or uncovered by sliding the tube. The instrument is handy, but only a very small amount of the lapis can be attached to the button, and hence only very small surfaces can be cauterized.

Rauchfuss has hit upon a method of construction which has already been attempted. His instrument is like the preceding. It has caps of various sizes, which may be screwed on; the end of the covering-tube, corresponding to the curvature, consists of a closely confined silver spiral wire, by which it is possible to give various curves to the instrument. The button of lapis melted on, may be enlarged by repeated immersion. As the lapis is brittle when it has once become damp the same button can only be used with safety for about an hour. Lewin makes use of a very similar porte-caustique. Both instruments have however the slight disadvantage, that the wire carrying the lapis button becomes warm by sliding, and hence the button readily rubs along the cup, and may thus be dislodged.

I hoped to be able to remedy this evil by applying to the wire within the cup, above the caustic, a contrivance which in the shape of a disc fills out the cup and prevents the wearing of the cauterizing button. The curved part of this porte-caustique consists of a very thin tube of hardened caoutchouc. Below, cups of various sizes may be applied, and brushes, and platinum wires with buttons may be fastened upon the wire. In order to cauterize larger surfaces, I have had made a pear-shaped button, which may be screwed on. This is dipped into melted nitrate of silver either wholly or only on one side as necessity may require; here I would make the observation, that the caustic adheres more surely to the button when the latter has been previously warmed. Charpie threads or small sponges for the application of liquid caustics, may be fastened to a wire handle which can be screwed on. The tube of caoutchouc may be curved to any degree after a slight previous warming. For the pharynx and epiglottis, where the cauterization is more easily applicable, a simple wire, with the button bearing the melted caustic, and a wooden handle, is sufficient. I am glad to coincide with Lewin on this point.

Caustics and astringents are applied in solution, and even in a concentrated solution, in order to avoid the frequent repetition of the operation, quite as frequently as in a solidified state. The phenomena are in general the same: great care must be taken that none of the liquid drops into the larynx or pharynx.

Dittel's apparatus for this purpose consists of a catheter tube open below, and having two rings on the external end by which to hold it. There is a movable pencil inside. Fournié has produced a forceps fashioned like a lithotrite, with which the epiglottis is to be seized and drawn forwards. In the arm which passes over the epiglottis, there is a very small tube, so that by means of a syringe introduced into it, an injection may be made into the larynx while the epiglottis is held fast; a sliding cautery might also be attached in place of the tube. This instrument, however, is clumsy. A proposal of Czermak's is worthy of notice; it is to combine the mirror with a syringe and to cause the syringe tube to open upon the surface of the mirror in a given direction, so that on emptying the syringe, the stream must always strike a given point, which is reflected on a particular spot in the mirror. The attempts with instruments of this kind have thus far succeeded favorably.

Trousseau long since introduced over the epiglottis a small tube filled with fluid, having first pressed the tongue down so that the apex of the epiglottis was visible; at an inspiration he allowed the remedy to fall in *guttatim.* Thompson in a similar manner applied a delicate shower-bath. The former has also given the best description of the phenomena which follow cauterization of the glottis. Others, like Green, *e. g.* would pass a wet sponge into the larynx without hesitation, just as they would into a chimney-flue, or they would press out a wet sponge upon the posterior wall of the pharynx during an inspiration. That these methods of proceeding must nearly always fail, seems almost certain. As has been already remarked, the author's porte-caustique may also be used as a pincette for the introduction of charpie, small sponges, or small bunches of raw silk.

A third method of applying remedies to the numerous membranes of the larynx and pharynx is in the form of powder.

This procedure has been long practised in so far as this, that powdered substances have been placed in little tubes, which were stuck into the patient's mouth; with the nostrils closed the patient inspired quickly. As the little tubes were open at both ends, the powder was inhaled and some tendency to cough was a sign that the operation had succeeded. Without the use of the mirror, this procedure is entirely wanting in system: if a small portion of the inhaled powder reaches the larynx, it is a mere chance. Störk makes use of a tube about five inches long, which anteriorly is beak-shaped, and posteriorly is attached to a flexible caoutchouc tube, the free end of which the physician takes in his mouth, and thus he is left entirely free in all his movements. The powder is placed in the horizontal part of the tube, so that none of it can fall out, and it is then emptied by blowing. Gilewsky has furnished this tube at its free extremity with an olive-shaped body, which is thickly perforated, and with a sliding fenestrum upon the back of the straight portion. The instrument of Rauchfuss is more convenient than the preceding. The tube in his is much thinner, and therefore interferes much less with the light. The curved portion may be removed, so that the operator is enabled to change the horizontal portion. The extremity of the beak has either a sieve-like perforated olive-shaped extremity, so that the powder may be scattered in all directions, or a movable button with a single lateral opening, so that the powder can only be projected in a single specified direction. The tube is held at its outer extremity by the index and middle fingers. The emptying of the tube is produced by pressure with the thumb upon a caoutchouc ball which is attached to the external extremity. Leiter has made this instrument for me out of hardened caoutchouc, a material which has been as yet by no means sufficiently esteemed in Akiology.

The insufflation must take place during an inspiration, otherwise it will be thrown back. If the vocal chords are to be reached in their greatest extent, the glottis must be closed during the insufflation, and according to what has already been said, this closure takes place by the production of an inspiratory tone. The sequelæ of insufflation are tickling, little if any tendency to cough, no suffocative paroxysms, no pain. Remedies soluble in water are to be preferred, as these act mechanically by contact, and afterwards chemically by solution.

Gilewsky wishes to limit insufflation to simple catarrhs of not too long standing, and thinks the effect is not so much a cutting short of the course of the disease, inasmuch as there is an increase of the catarrhal symptoms, but rather a special curative alteration of the tissues. Lewin rejects insufflations because it is difficult to reach definite points, and declares himself especially opposed to the nitrate of silver in powder, because several times after its use he saw blood coughed up.

Hitherto the remedies used for insufflation have been very finely powdered sugar, alum (alone or with sugar), and nitrate of silver; it is usual to combine the latter with the sugar of milk in the proportion of 1 : 10.

Still another mode of applying pulverized remedies, is the inhalation of liquids containing a fine dust, by means of various well-known apparatus which exist in various shapes, from that of Sales-Girons down to the simple and well adapted one of Schnitzler.* I may take it for granted that these are known to the readers of these pages. After numerous efforts, there is now no longer a doubt that pulverized solutions may in this way be introduced into the larynx and the trachea, even when these organs are diseased. Thus is established the great importance and future usefulness of this method for the local treatment of the diseases which now engage our attention; valuable observations upon this method already lie before us. In general it is known that simple acute inflammations frequently pursue a favorable course by an appropriate régime without other treatment, and that chronic forms often present the negative results of treatment. In the latter, according to my own experience, the success of local treatment is sometimes manifestly increased by the external application of croton oil; but I cannot forbear

---

* Gibb has devised a very simple and useful instrument for this purpose; he uses it very extensively and to good advantage. It is simply a caoutchouc bag attached to a silver or gold tube curved somewhat like a catheter; at the end of a tube is a platinum capsule with very fine perforations, "invisible to the naked eye." (TR.)

in this connexion to condemn unequivocally the frequently purposeless application of croton oil and tartar emetic ointment, in the use of which such sad abuse is perpetrated. Next to some other remedies, belladonna and the succinate of ammonia are particularly in repute in diseases of the larynx and of the pharynx. Locally, in chronic cases, iodized glycerine has been applied with good success in various degrees of strength; so also the bromide of potassium.

Diseases of the Cartilage and Perichondrium.

Under this head, with the exception of a few cases, we have only to consider the inflammation of the perichondrium, which sometimes pursues its course independently, and sometimes leads to alterations and diseases of the cartilage itself.

Perichondritis occurs most frequently upon the arytenoid cartilages, then upon the cricoid, and rarely upon the thyroid cartilage. It exists either primarily, or occurs secondarily by a continuation of the inflammation from the sub-mucous tissue, in tuberculosis, syphilis, and typhus, and this secondary form may occur in a two-fold manner, inasmuch as the perichondrium itself may be first diseased ; or diphtheritic ulcers, progressing still deeper in their destructive course, may at last reach the perichondrium, and lead to its disease: thus large collections of pus and ichor are formed, in which pieces of cartilage and even the entire arytenoid cartilages lie free.

The product of perichondritis is either pus (in other words, laryngeal abscess), or cartilage or bone substance, more frequently the latter, and then attended with increase of size of the parts and constriction of the tube. This product may be deposited upon one or both surfaces, and causes, consequently, corresponding alterations in the superior layers, extending itself, according to the locality, in various directions. If pus is accumulated upon the inner side, it forms protruding and constricting swellings which can be recognised with the mirror, which demand opening, and which often prove indications of laryngotomy. At last it bursts and pus is poured out, being as likely to induce suffocation as it is to be expectorated ; in the latter case, the loosened bits of cartilage are cut off with it. These pieces may also be discharged afterwards; even long after a successful tracheotomy they may yet stop up the canula, and thus lead to death by suffocation: examples of this kind are on record. Or the pus is deposited upon the external surface, and finally discharged externally or through neighboring cavities and passages. The destruction of the cartilage occurs either from absorption or after previous ossification from caries and necrosis. But few observations upon this subject are recorded; Türck has published some. Thus five of them are subsequent to typhus.

There was inflammatory swelling of the vocal and ventricular chords; inflammatory tumefaction or œdema of the mucous membrane of the arytenoid cartlages; and œdema of the vocal chords, the latter accompanied by the severest dyspnœa. In all of these cases there was coughing, alteration of the voice, pain, dyspnœa, and at times dysphagia. All five underwent tracheotomy; two died, three recovered but with permanent constriction.

Türck had one patient, thirty-four years of age, who had suffered for eight or ten days with hoarseness, pain in the larynx, and dyspnœa; by means of the mirror he discovered that the left vocal chord was very much bent forwards, immovable, with its edge reaching beyond the middle line, without inflammatory appearances, and of a normal color. So also the left Santorini and arytenoid cartilages were immovable, and their mucous membrane was swollen. The left laryngo-pharyngeal fossa (Sinus pyriformis, TORTUAL) was enlarged. The right vocal chord, as well as the movement of the right arytenoid cartilage, was normal. The patient died the following night. The post-mortem examination showed upon the left half of the cricoid cartilage an abscess as large as a hazel-nut, which extended partly beneath the plate of the thyroid cartilage outwards, partly into the sinus pyriformis, partly beneath the undermined left vocal chord into the cavity of the larynx, and thus reduced this to a simple chink extending from before backwards. The left half of the cricoid cartilage was partly denuded of perichondrium, and upon a small spot in its posterior section it was rough and infiltrated with tubercles. There was also tuberculosis of the pleura. Türck considers the symptoms given above as characteristic of an abscess which has discharged itself.

The author and others also have observed cases of perichondritis and of necrosis of the cartilage, but there is nothing of special importance to be added to the above.

The chief therapeutical indications are the early opening of the abscess, tracheotomy in threatening danger from suffocation, and afterwards careful watching of the patient for a long time. In such cases repeated examinations with the mirror from above, and by all means also from below, are of the greatest importance. An abortive treatment of perichondritis is scarcely to be thought of.

Inflammation of the Pharynx.

I consider it advisable to say something upon this point, because these diseases of the throat, so called, often give occasion to examinations with the mirror. On the one hand they are not sufficiently regarded by the physician, and on the other they are very annoying to patients, if we can place confidence in their reports, and can judge from all they will endure to be freed from their suffering.

By the word pharynx we comprehend here the velum palati, the posterior wall of the cavity of the pharynx down to the œsophagus and to the larynx, its lateral walls (inclusive of the region adjacent to the opening of the Eustachian tube), the root of the tongue, and the space between the tongue and the epiglottis. The patients complain of dysphagia, especially in swallowing dry, spicy, hot, or acrid food ; the saliva collects to excess in the mouth ; the breathing may be rendered difficult in consequence of tumefaction or from exertion ; attacks of congestion of the head, and even of suffocation occur, especially in the night, when the mucous membrane becomes dry from the mouth remaining open ; the tone of the voice becomes harsh, rough, hoarse, and unmetallic, especially after continued speaking or singing. The subjective phenomena are, a feeling of pressure or of fulness, or that of a foreign body, which is stuck at some point or other, and can neither be raised up nor swallowed ; a drawing and tearing sensation in swallowing and speaking, especially in the morning.

The seat and cause of these troubles are not always to be discovered by a simple inspection. With regard to the origin and first cause of the disease, there is seldom anything sufficient to be ascertained, as the development of the trouble proceeds very gradually, so that as a rule, only the forms which have become chronic are presented for observation, and these, therefore, will chiefly occupy our attention. Acknowledged causes are, colds, acrid irritating food and drinks, foreign bodies which are stuck fast, as fish-bones and splinters of bone, continued overexertion of the vocal organs, constitutional diseases, as syphilis, scrofula, mercurialismus, etc. Frequently there are combined with it catarrh of the nasal and lachrymal passages, of the larynx, and of the Eustachian tube. The prognosis is in so far unfavorable,

9

as chronic inflammation of the pharynx frequently offers to curative efforts an opposition not to be overcome. The destruction
of individual organs, with permanent trouble, or dangerous accidents resulting from a secondary œdema of the entrance to the
larynx, etc., occur indeed very rarely in connexion with the
uncomplicated forms.

Various forms are described according to the extension of
the inflammation to the various tissues and the alterations of the
same. If the disease actually affects the papillary bodies and the
mucous glands, there is an increased secretion, and the coloring
is at first fresh red, and afterwards bluish, either evenly spread
over the surface or in spots and flecks. The mucous glands are
projecting, resembling gravel, swollen, reddened, or filled with a
yellow infiltration (follicular or granular catarrh); there are erosions, granulations, small blisters, and superficial ulcerations
(pharyngitis ulcerosa), which are no more to be always attributed
to syphilis than are scars in the wall of the pharynx; there are
also aphthæ. Sometimes the diseased surface is covered with a
visible layer of thin mucus, or there are found enlarged tortuous
veins (pharyngitis varicosa), the rupture of which may give rise
to bleeding.

Lewin describes such a case in which the expectoration of blood for
many years, accompanied with a short cough, gave rise to the most obstinate
hypochondriasis, in spite of the assurances of the physicians, that the patient's
lungs were entirely sound.

Frequently there is thickening of the mucous membrane and
of the sub-mucous tissue from œdema and hypertrophy, hyper-
plastic pharyngitis.

Lewin describes a case of this kind in a strong man forty-one years of
age. After previous over-exertion, the voice had been for two years hoarse,
rough, and unmetallic, and always became more hollow after special efforts.
Frequently paroxysms of dyspnœa occurred, and terror at night from dryness of the throat. The mucous membrane of the posterior wall of the pharynx was so much swollen that opposite to the apex of the epiglottis it
formed a protrusion excavated like a crescent, into the cavity of which the
edge of the epiglottis was thrust, so that between the two there was left
only a fissure of about two millimètres (0.07 inches) breadth. The examination of the larynx was difficult in consequence of this constriction.
The root of the tongue was fissured both horizontally and longitudinally,
and was covered with a thickish grey deposit. The mucous membrane of

the posterior nares was swollen, the posterior surface of the soft palate was flecked, red, and yellow. Improvement.

The author has a similar case to record. A young man complained of suffocative paroxysms, which for several weeks had occurred frequently and suddenly, and which placed all in the vicinity of the patient in mortal fear. It was observed that before the attacks the breathing was rough and stertorous. If the patient was aroused, he felt great dryness, heat, and tension in the pharynx, all of which feelings passed away after swallowing a little mucilaginous drink; he could then sleep again for some time. These attacks occurred more frequently after having taken cold. His voice had a peculiar short, dry, and empty sound, and his speech was in short sentences, and evidently with foresight and attention. The patient stated that he could not swallow large or dry morsels of food. In the upper part of the pharynx only a few dilated veins could be seen. The mirror showed that the larynx was normal, the fossa between the gums and the pharynx reddened; the posterior region could not be seen in consequence of a painless red shiny tumor, having a horizontal position, which extended into the lower posterior region of the pharynx. After repeated painting with a caustic solution of iodium and glycerine, the attacks ceased, and the tumor became smaller.

The uvula may become so much enlarged by hypertrophy and œdema, that it continually touches and irritates the tongue or the epiglottis; such a condition is sufficient indication for cutting off the uvula. Such a case of acute œdema of the uvula, with extravasation of blood, by which it became as large as one's thumb, I saw in consequence of having seized the uvula with a wire sling for the purpose of facilitating the examination. Scarification and an ice-water gargle relieved it quickly.

The mucous glands at the root of the tongue, and the glosso-epiglottic ligaments, undergo hypertrophy in chronic catarrh, in syphilis, and tuberculosis. At these points, and in the glosso-epiglottic sinus, there may be ulceration and the formation of out-growths.

Cases in which a catarrh of the pharynx has extended to the Eustachian tube, accompanied by deafness, pain in swallowing, &c., have been observed by Czermak, Lewin, Störk, and the author.

I wish to allude still farther to a case of granular pharyngitis, which bids defiance to all curative efforts, even the most intense cauterizations. Of late I have risen to three hundred inhalations a day of a pulverized solution of tannin, in the proportion of 1 : 100. The result is still uncertain. The entire cavity of the pharynx is diseased; the mucous membrane is covered with an even layer of thin mucus, and it is so irritable that it becomes reddened during the examination in consequence of the tension; indeed, I can tell by the more marked injection whether the patient has eaten his breakfast or not. Here the mucus collects, particularly in the fossa between the gums and the pharynx, and upon the epiglottis; it produces a rattling sound in breathing, and is with difficulty ejected. By extension of the disease there has been developed a chronic periostitis of the os hyoides, known by the pain on pressure and swallowing, by thickening of the body of the bone, and by an occasionally recurring feel-ing of heat and tenderness in its vicinity. Upon the use of strongly astringent gargles, the mucus coagulates in little white lumps, like albumen, and is then easily expectorated. In the same way, after the inhalation of tannin, flecks of coagulated mucus lie in the pharyngeal parts.

The treatment has for its first object the avoiding of all injurious things. Cold, in the form of ice-pills and gargles, does good service. Störk, I think, first applied to a catarrh of the pharynx cold water as a fine shower-bath by means of a ball of caoutchouc, and with good success. Then follow the astringent gargles, of various sorts, only they must be comparatively strong if any effect is to be derived from them. I would espe-cially recommend the tinctura opii composita and that of thuja occidentalis. As a last resource remains cauterization; fre-quently a simple touching does not suffice, and the granulations have to be destroyed mechanically, as in trachoma. This proce-dure is quite difficult to accomplish, in consequence of the mobi-lity of some of the structures of the larynx, and of the vomiting which may be induced. I have applied nitrate of silver, sul-phate of copper, a caustic solution of iodine and glycerine, but in some cases I have accomplished nothing, unfortunately.

Latterly I employ the inhalation of dissolved powders, in the form of spray, proceeding on the supposition that thus we can reach all parts of the cavity of the pharynx, and can supply by a continuation of the impression that which is wanting in energy; I allow the patient to breathe very gently when I do not wish much of the remedy to reach the air-passages. The results hitherto encourage one to the further prosecution of this method.

Diseases of the Nerves and Muscles of the Larynx.

Disturbances of the Sensibility.

Concerning these forms of disease, of which hyperæsthesia occurs the most frequently, there is in general but little to say, and from our stand-point only this, that the laryngeal mirror in some such cases is of use inproving the absence of organic alterations; to be sure, anæmia of the mucous membrane may be observed in such cases.

Mandl recommends cold applications, and at all events slight cauterization with a weak solution of nitrate of silver.

Disturbances of Motility.

Here it is scarcely possible to separate the diseases of the muscles and nerves. There is also nothing to report from a laryngoscopic point of view upon the spasmodic muscular affections of the larynx—inasmuch as in those cases in which the examinations have been made, the structures of the larynx have been found normal.

Mandl (Schmidt's Jahrbücher und Gaz. des Hôpitaux) says: the negative result of the physical exploration, the absence of expectoration, the character of the cough, the periodicity of the attacks, &c., facilitate the diagnosis. If these attacks become convulsive, they form a kind of laryngeal chorea. The same often occurs in connexion with acute affections of the air-passages. These attacks should not be confounded with the rough spasmodic barking which occurs frequently as a concomitant in hysterical or epileptic cases. The more developed types of this affection constitute the epilepsia laryngea, which is curable by tracheotomy (?). In these neurotic affections, sedative, narcotic, and anti-spasmodic remedies suffice, of which latterly the valerianate of atropine has been used with success by Dr. Lac. Mandl found in some such cases that painting the posterior wall of the pharynx and the epiglottis with a solution of the iodide of potassium was serviceable.   Türck reports one case which he is inclined to designate as a cramp of the crico-thyroid muscles.

I have myself made examinations in whooping-cough, and in some other cases where there were coughs of a very peculiar nature, without obtaining any result.   I shall speak of the paralytic conditions of the larynx in the sections upon the alterations of the voice.

Neoplasms in the Larynx.

This division constitutes one of the most brilliant features in laryngoscopy, and records an actual progress in the medicine of our time.

Besides the individual observations of Czermak, Türck, Störk, Gerhardt, Moura (Fauvel),* we have a large and able treatise of Lewin's in the *Deutsche Klinik*, which at the time of our going to press is not yet completed.

The History of Laryngeal Polypi, by Ehrmann, published in 1851, records thirty-one cases of morbid growths in the larynx, two of which were Ehrmann's own observations.   From Rokitansky's observations, published in 1852, there were ten more cases added.   The compilation of Middeldorpff, in his " *Galvano Kaustik*" published in 1854, comprises a single observation of his own and sixty-four cases taken from the literature of this subject.   Lewin has up to this time collected about eighty cases from the latter source.

The greater number of these neoplasms were first discovered upon the dissecting-table.   In only three out of the sixty-five cases was an operation attended with success; this arose chiefly from the uncertainty of the diagnosis; for sometimes no operation was performed when the circumstances were favorable, and again patients were operated upon under unfavorable conditions.   All the signs given as proofs of the existence of morbid growths in the larynx are uncertain; they may be altogether wanting in certain positions and dimensions of the formation, or they may be masked by other diseases.   Such signs are, alterations of the voice, coughing, dyspnœa, an asymmetrical condition of the

* And the Author.   See Appendix.   (Tr.)

larynx, disagreeable sensations, auscultatory signs, especially the
so-called valvular murmur, which, however, as it now appears,
is infinitely rare. Lewin advances as diagnostic of the seat
of these growths a sign which has lost greatly in value
by the introduction of the mirror, viz. that the inspiration will
be audible, rattling, whistling, or sawing, &c., and the expiration
more or less free if the polyp has its seat above the vocal chords,
and *vice versâ*. Even the most valuable objective sign, viz. the
expulsion of portions of · the tumors, with the simultaneous
existence of the other phenomena, leaves us in the dark with
regard to the seat and the extent of the formation. All possible
diagnostic errors were made; all manner of silly, purposeless,
expensive, and even injurious methods of cure, were introduced
before the laryngeal mirror placed the means in our hand of
making an exact diagnosis, and that, too, at a time when there
is no danger in delay, and when the necessary operative attempts
may be quietly considered. Among the eighty cases cited, a
laryngeal tumor was only six times with probability diagnosed,
and hence an operation was abstained from or attempted too
late; in one case, where the result seemed to promise favorably,
the canula was withdrawn too soon, and thus the death of the
patient was induced.

The history of these diseases has experienced an entire revo-
lution from the introduction of the mirror. Rightly does Lewin
say, that a tumor in the larynx should not now be so certain a
cause of death. It may happen that in some cases the examination is entirely impossible, but then the proper moment has been
allowed to pass by, for tumors of the larynx are not wont to
grow rapidly. Since the introduction of the laryngoscope,
Czermak, Türck, *et al.*, have diagnosed about thirty, and Lewin
alone about fifty growths in the larynx. This frequency may
naturally astonish us. But we should remember that many of
the observations which are on record prior to the days of laryngoscopy, were first made by chance at the dissecting-table,
and that on the other hand in many cases of sudden death,
the tumors existing in the larynx were not observed, or the
larynx itself was not sufficiently examined; for as false dia-
gnoses were made in almost all cases, we can easily suppose that
the dissector also proceeded upon an erroneous course, and the
consequences of the stenosis of the larynx would be taken as

the primary cause of death. Since the introduction of the laryngoscope about 100 cases have become known.

The operative procedures which were in use before the days of the mirror were as follows:

1. Removal through the cavity of the mouth, of which there are two cases, one by Koderik, by means of a flexible instrument like a string of beads; nothing further is known of this case; and the second by Middeldorpff, by means of a galvano-caustic écraseur; the tumor could be seen upon and behind the epiglottis when the latter was made visible by the mouth being opened and the tongue being hollowed.

2. Artificial opening of the air passages, and excision of the tumor. This procedure was proposed by Ehrmann, and by him applied but once. Although the probability of a correct diagnosis of a laryngeal polyp existed, and although small portions of it were coughed up, still the operation was only performed when danger of suffocation suddenly set in. After the opening of the larynx, and the division of the cricoid cartilage, and the first two tracheal rings, a canula was introduced; and then after the patient had enjoyed a respite of forty-eight hours, the larynx was divided in the middle line as far up as the os hyoides, while the canula was still allowed to remain. When the two halves of the thyroid cartilage were held apart, the growth was seen seated upon the left vocal chord, and was seized and removed with a scalpel. The patient soon recovered, and on the twenty-first day the wound had healed but the aphonia remained; six months afterwards the patient died from typhus.*

Ehrmann attaches particular importance to the direction that tracheotomy should be performed some time before laryngotomy and extirpation, that thus the patient may recover a little from the impression, that the passage of the blood into the trachea may be hindered, and that the larynx may be carefully examined. But the shock is always quite considerable, and the voice may be irretrievably lost from injury of the vocal chords, or of the nerves and muscles.

* Through the kindness of Dr. H. B. Sands of New York, the translator obtained access to the records of a case in St. Luke's Hospital, in which Dr. S. opened the larynx for the removal of a tumor. By the aid of the laryngoscope a small tumor was discovered projecting from the left ventricle of the larynx. The air-passage was laid open from the second or third tracheal ring up, and the thyro-hyoid membrane was also divided, together with the base of the epiglottis. A round fleshy excrescence, as large as the end of the little finger, projected rather more than a quarter of an inch into the cavity of the larynx, and concealed the vocal chords. It was cut off down to the level of the chords with a pair of curved scissors; but it was impossible to effect its complete removal owing to its firm and broad attachment to the wall of the ventricle. The actual cautery was therefore applied. Complete closure of the opening was effected by the end of the fourth week. The microscopic examination showed the tumor to be of a cancerous nature, consisting exclusively of nucleated cells (epithelioma?). The voice improved very much after the operation, and six months afterwards there was no appearance of a recurrence, although the voice was a little rough.

This method, as well as that of the écraseur, should not be applied when it is simply an alteration of the voice that we have to deal with. But the performance of Ehrmann's operation would only be facilitated, if the operator had learnt by a previous examination with the mirror, the seat and size of the tumor.

3. Laryngotomy below the hyoid bone, by Vidal, Malgaigne. The entrance should be made parallel with the lower border of the lingual bone. The remark of Hyrtl, that thus we come directly into the pharynx while the larynx is not accessible, is not altogether correct. To be sure we are above the base of the epiglottis, but still we look down upon the arytenoid cartilages, and the posterior portion of the ary-epiglottic folds. If, then, according to Malgaigne, the epiglottis is drawn forward into the wound, by seizing firmly hold of its apex, the entrance to the larynx becomes more freely accessible. But it is true that the cut may be carried into the base of the epiglottis, as its attachment is not always the same, and we cannot sufficiently determine this point by inspection, or by the touch. By this accident the purpose of the operation would be frustrated, the inspection of the larynx rendered impossible, and the injury would be very severe, while otherwise it is not at all severe. The superior laryngeal artery and nerve run upon the upper border of the thyroid cartilage, and would not therefore be injured if the operator kept close to the edge of the lingual bone.

This operation seems to promise well for laryngoscopy. I brought these views forward three years ago in a memoir which was never published, and now I endorse Lewin's views. The operation would be indicated in the case of growths between the tongue and the epiglottis, upon the latter, and especially in the entrance to the larynx.

If the cut was made more to one side instead of in the middle, it would, indeed, be somewhat more difficult to avoid injuring the superior laryngeal artery and nerve, but in this way, by a cut from one to one and a quarter inches in length, we could gain a fine view of, and ready access to, the ventricular and vocal chord of the opposite side. It should be mentioned that Vidal, when he brought forward the above operation, did not think of laryngeal polypi, but only recommended it for those abscesses and infiltrations of the spongy tissue at the base of the epiglottis, which obstruct the breathing. The objections, moreover, which were brought against Vidal's operation, were completely removed by the fortunate result in a case reported by Pratt, who was the first and only one to carry out this operation upon the living subject.

A tumor, which had caused great difficulty in swallowing and breathing, could not be seized from the pharynx. Pratt performed subhyoidean laryngotomy; the growth was seated upon the left half of the posterior surface of the epiglottis, and protruded into the pharynx. It was removed, and proved to be firm and fibrous, and of a greyish-white color. No vessel was ligated. The wound was united by three strips, and healed well. All the difficulty was entirely removed.

To these operations there are a few to be added, since the introduction of the laryngeal mirror, which could only have been performed by its aid; and this is a positive advantage of laryngoscopy, that even a direct operative procedure, by its assistance, is made possible with greater certainty, although also with greater difficulty. All these methods of laryngoscopic surgery may, if they are conducted with delicacy and knowledge, yield every advantage to the patient, and but very slight disadvantages.

These various methods are *cauterization* by means of the instruments described in a previous section. It is specially adapted to small soft structures, or to such as have a broad basis which cannot be removed entirely nor even piecemeal; or, finally, to cases where considerable hæmorrhage is to be feared. I would call attention to the fact, that it is not possible in the larynx to cauterize as severely, or to allow the cautery to work mechanically, as we do elsewhere when we wish to destroy the parts, because there is not sufficient time, and the individual parts are too movable; and a superficial cauterization in many cases will lead rather to a more rapid growth than to a shrinking of the tumor; for we apply the pencil of nitrate of silver in a very different manner when we wish to destroy the hardened edges of an ulcer, and when we would excite indolent granulations.

Extirpation. For this purpose we employ curved forceps, like the pharyngeal forceps, which serve for grasping and tearing off (Lewin); also, two-bladed forceps, the blades of which are separated from each other by a spring movement, like Hunter's lithotomy forceps, and which are closed by sliding forwards a tube which covers them. Such instruments have been devised by Lindwurm and Lewin. Also, scissors curved on the edge; the lock is at the curvature. Lewin placed upon each blade of the scissors near its point, two sharp teeth, which, upon cutting, shut into each other, and would penetrate the part to be excised, and thus prevent its falling down into the air-passage; he does not agree with Bruns in the opinion that this accident would be of but little consequence, and that such a structure would be coughed up or absorbed. Bruns has invented a pincette with small cutting blades on the extremities of the arms, an exceedingly ingenious instrument, for the purpose of cutting off morbid growths, or at least of cutting into them and thus preventing their nutrition. Moura has in two cases of epithelioma introduced curved metallic sounds into the anterior extremity of the glottis, seeking to destroy these growths mechanically by friction. The result was, that portions were several times cast off, and the patients experienced some alleviation. I am inclined to think, that if it were merely a question of destroying the nutrition of tumors, this could be more simply accomplished by crushing them with the forceps than by the use of cutting instruments, since with the former, the possibility exists of tearing away entirely, or at least partially, the part seized. It is to be understood that all these instruments must be properly curved, and that one curvature will not be sufficient for various cases.

All these operations are in many ways wearisome and difficult, more for the operator than for the patient. None of these operations have as yet been performed under narcosis. Lewin rightly recommends as useful a slight fixation of the patient's head, and employs for this purpose a standard similar to that which photographers use. By this means the patients are spared much of that annoyance and irritation which in Bruns's cases proved so uncomfortable. The outstretched tongue must also be firmly held, and this the patient will best accomplish himself, by means of his own hand and a napkin. Every instrument for this purpose Lewin throws aside, and we agree with him. In such cases for operation the use of some method for fixing the laryngeal mirror, or for elevating the epiglottis, may be of great use. That in many cases the patients must be, so to speak, previously educated for the operation, every one will understand.

Lewin's compilation gives as the seat of these morbid growths,

I. The epiglottis in twenty-three cases, in some exclusively, in others combined with growths upon other parts of the larynx. Two of these cases are laryngoscopic. These formations were seated nineteen times upon the laryngeal surface, and in the majority of cases upon the lower half, three times upon the lingual surface. In so far as the histologic character of these tumors is to be discovered from the description, they were carcinoma, eight times (?); fibroplastic tumors, four times; epithelioma, five times; polypi (?), four times; hydatids, once; syphilitic growths, twice. Why the lower surface of the epiglottis is peculiarly the seat of these growths, cannot be explained.

II. The ary-epiglottic folds appear exclusively or in combination, as the seat of morbid growths in nine cases. Of these four can be classed as carcinoma, one as epithelioma, two as of connective tissue, one as sarcoma, and one as an enlarged lymphatic gland. One case is laryngoscopic.

III. Morbid growths were found in the ventricles of Morgagni twenty-one times. These were seated upon both sides five times; on the right side, five times; on the left, seven. In four cases the position is not given with any greater precision. They are, however, in regard to histology, classed as follows:—Cancroid, five; polypi, six; fibroid, four; epitheliom, two; hydatid, one. As circumstances which might favor growths in this region, are mentioned the frequent catarrhal affections and the abundance of glands in the ventricles. Two cases are laryngoscopic.

IV. Growths were observed upon the vocal chords thirty-three times. They were seated upon both vocal chords, eight times; upon the left, seven; and upon the right, sixteen times. Histologically, so far as their nature is given, they were polypi, without farther description, eighteen times; epithelial growths, five; fibroplastic tumors, three; cancroid, three; encephaloid, one; syphilitic excrescences, one. Here Lewin mentions also the observations upon the broad condylomata upon both vocal chords, which were found in so many cases by Gerhardt and Roth (see Syphilis). It is difficult to reconcile the frequency of these growths upon the vocal chords with the structure of the latter. Of the thirty-three cases cited, seventeen were laryngoscopic.

V. The ventricular chords are reported as the seat of these growths five times; and of these, three times the left, once the right, and once both. Two cases were laryngoscopic. These were more particularly described as growths of connective tissue twice; as a pediculated polyp once; and as epithelial cancer once. It is remarked that many formations which are described as proceeding from the ventricles of Morgagni, should more properly be attributed to the inferior surface of the ventricular chords.

VI. These growths were found upon the arytenoid cartilages three times, and all of them of connective tissue. Two cases were laryngoscopic.

VII. The anterior wall of the larynx appears as the source whence growths proceeded eight times, five of which were laryngoscopic; and

VIII. The posterior wall twice, with one laryngoscopic case. Lewin asserts that this extremely small number may find its explanation in the fact, that growths existing on this surface, which by respiration and utterance is so constantly stretched, and pulled, and pressed upon, in so far as they are not of a firm texture, must become destroyed in consequence of these mechanical forces, and thus give place to ulcerations.

IX. Lewin reports eleven cases of morbid growths below the vocal chords and in the trachea. These eleven cases comprise two pediculated polypi, three fibroids, three carcinoma, one lipoma, and two which were of uncertain character. Four cases were laryngoscopic. In conclusion Lewin collects :—

X. Fourteen cases in which the entire larynx and a greater or less portion of the trachea were the seat of morbid growths. All these cases, with one exception, were of a cancerous formation, and to classify them more minutely, epithelial cancer, seven times; cancroid, twice; the papillary form, once; and encephaloid, once. That carcinomatous masses should occupy a greater surface is not remarkable, and this fact might furnish an additional aid in diagnosis.

Besides these, there are ten cases of morbid growth in the air-passages, the nature of which cannot be determined from the reports; these embrace apparently 6 polypi, 2 cancroid, 2 osseous tumors.

Upon the whole, the most numerous are carcinoma, epithelioma, and tumors of connective tissue; 28 formations are described as pediculated polypi—18 of which were seated upon the vocal chords.

The total number in Lewin's report is rather too large, as some cases are recorded twice, partly because the formations were seated upon more than one point, and partly because the same case has been observed and described by two observers, e. g. one case by Türck, and also by Störk. Lewin has also included some cases, which may indeed be classed under the head of neoplasms, but which are not such in a surgical sense, for in this sense it is necessary that the neoplastic mass should be more or less separate in the form of a tumor, from the tissue in which it is placed. Thus, for example, I should be inclined to exclude from the neoplasms that case of Czermak's, of scrofulous induration of the sub-glottic mucous membrane and of the cellular tissue (see the section, examination through the canula); in the same way, too, certain cases of small nodules as large as a hemp-seed, which are only indications of granulations or of exuberant ulcers. And no less would I exclude Türck's case of circular constriction of the larynx, below the glottis, by means of a tumor (inflammation of the mucous membrane), and also the cases of constriction from projections produced by goître or by cancer of the œsophagus.

Still it should be mentioned that among the cases collected by Lewin, 129 in number, 36 were diagnosed by the laryngoscope. And as Middeldorpff in 1854 already noted 65 cases, we now know of about 100 additional which have been brought to light by the laryngoscope. Surely a satisfactory result for four years!

I append some observations made since Lewin's work was published. Thus Bruns records the following case in his *Brochure:*

The case was that of a brother of the worthy Professor. The tumor, a fibroid, sate imbedded in a fold of mucous membrane beneath the edge of the left vocal chord, and was quite movable in its entire length. By means of the cutting-pincette devised by himself (see above), Bruns, after seizing the tumor repeatedly, made several cuts in it, which were not felt at all by the patient. The hæmorrhage in the first sitting amounted to about two ounces in three-fourths of an hour. The tumor was chiefly of connective tissue; if one could judge from its condition, it was becoming gangrenous. The general reaction from the operation was quite considerable, and from the commencement of the operation, its course very rapid (a few days only). The success was complete.

Türck states in a paper upon papilloma and other small excrescences, that these are not unfrequently the products of chronic inflammation, and that the granulations of ulcers, and the various mucous enlargements and elongations upon their edges, may be taken for neoplasms; hence in such cases we must decide with caution. Referring the reader to some of my previous observations, I coincide with this view of Türck's.

Türck relates one of these cases, in which small pale pointed growths were seated upon the lower section of the posterior surface of the epiglottis, and upon the anterior portion of the left vocal chord. The accompanying inflammatory enlargement of the entrance to the larynx, led to tracheotomy. The examination through the canula showed similar growths upon the lower surface of the anterior half of the left vocal chord, and upon the anterior wall of the larynx. Very probably syphilis was the cause of it all.

The same author also reports an observation of a body as hard as cartilage, larger than a pea, with a broad base, proceeding from the ventricular chord, or from the ary-epiglottic fold upon the left side; and a second case of a similar uneven body larger than a hazel-nut, which had a broad attachment to the posterior surface of the epiglottis and to the left wall of the entrance to the larynx; and finally three cases of cancer of the larynx, all of which were on the right half of the epiglottis and upon the right arytenoid cartilage, as well as upon the right wall of the pharynx, accompanied by hoarseness, bloody discharge, and by a disagreeable odor of the breath. In all three cases, the loss of substance upon the epiglottis amounted to perforation.

In another place Türck describes three laryngoscopic cases of tracheal tumors; two were strictures of large extent, one of them from cancer; in the third there was a small round tumor upon the epiglottis, and one also upon the superior portion of the posterior wall of the trachea accompanied by catarrh.

The following cases already reported by Lewin belong here.

1. Epitheliom upon the left vocal chord, aphonia, dyspnœa, wearisome cough, misapprehension of the disease for many years; ecchymoses upon the root of the tongue. The structure had a broad base, covered the larger part of the left vocal chord, and only allowed the posterior portion to be free. It was 12-14 millimètres long (0.47—0.55 inches), and resembled in size, form, and appearance, a good-sized raspberry. Extirpation after several sittings; about an eighth of it was left behind. The voice became much better; the annoying symptoms vanished.

2. A girl 15 years old, hoarse since childhood, at last aphonic. Slight dyspnœa; misapprehension for many years. There was a two-lobed polyp upon the anterior insertion of the vocal chords. Complete extirpation in several sittings. Six pieces were removed; cauterization of the remnant; the voice after the operation was resonant; at first still some impure tones. The individual fragments were from the size of a lentil to that of a bean; soft, granular. The microscope showed epithelial scales, and a delicate structure of connective tissue.

3. Aphonia for three years; dyspnœa; the feeling of a foreign body in the larynx. A neoplasm in both of the ventricles of Morgagni, and upon the anterior attachment of the vocal chords, partly pediculated, and partly with a broad base; at first cauterization, by which some portions were coughed up. Extirpation with the forceps; the voice gradually more and more clear. A mucous polyp, epithelial cells, and very vascular connective tissue. Numerous small ramifying glandular sinuses; atheromatous hardening of the vessels.

4. A man 56 years old. Three weeks before, a fish-bone had lodged in his throat; at first there were suffocative paroxysms in attempting to speak; finally hoarseness remained, and a sensation as if a soft cork stuck in the throat. Ulcerative pharyngitis, enlargement of the mucous glands at the

root of the tongue. The vocal chords were a bright red, and injected; a soft mass was growing out from the right ventricle of Morgagni, and was gradually increasing. Upon the application of the nitrate of silver *in substance,* the tumor vanished in nine days. The voice was at the commencement of the treatment sometimes more hoarse, as the chords were occasionally touched, but afterwards it became clear and resonant.

5. Burning sensation in the throat, hoarseness and cough, frequent hawking, loss of voice. Pharyngo-laryngitis. Below the anterior attachment of the vocal chords, a polyp as large as a grain of coffee, long and with a thin pedicle. Cauterization with a concentrated solution of nitrate of silver; voice good after three weeks.

6. Pressure, tickling, burning in the larynx, sometimes the sensation as of a foreign body. Below the anterior attachment of the vocal chords, an oblong growth of about 4-6 millimètres (0.15—0.23 inches) in length, with a thin pedicle which was thrown by sudden expiration into the glottis. Cure by cauterization.

7. Numerous excrescences upon the posterior laryngeal wall. The breathing was rough, since the occurrence of attacks of croup, 2¼ years before. For the ten weeks preceding, dyspnœa and suffocative attacks, slight chloroform narcosis, (it was a child.) Cauterization with a solution of nitrate of silver, *partes æquales,* applied with a camel's hair pencil. Coughing, bloody mucus. Disappearance of the phenomena.

8. A boy seven years old. Hoarseness and dyspnœa since an attack of erysipelas of the face two years before. Laryngo-pharyngitis. Neoplasm proceeding from the right ventricle of Morgagni. Cauterization with nitrate of silver. The structure was quite firm.

9. Hoarseness for thirteen years. Voice hollow and rough. An oval polyp behind the anterior attachment of the vocal chords.

10. Loss of voice for a long time, without any apparent cause. On both vocal chords, especially anteriorly, papillary growths hanging into the glottis. Some were extirpated, and consisted of pavement epithelium, with a central kernel of connective tissue. On the posterior wall, just above the vocal chords, there was a conical-shaped neoplasm standing upright: in utterance it came to lie upon the posterior portion of the glottis.

My own observations upon neoplasms in the larynx are not very numerous. Besides two cases of growths of connective tissue upon the anterior angle of the glottis, accompanied by chronic catarrh (tuberculosis), I have repeatedly seen Czermak's first case of neoplasm, and I can assert that the formation, which I consider as a product of a chronic catarrh, has not altered since the spring of 1859. I have also seen in Paris, at M. Moura's, the two cases of epitheliom reported by Fauvel.

Recently two cases have come under my own observation. One was a merchant in Upper Austria, who has been voiceless for some time. Growing out of the right ventricle of Morgagni there is a structure as large as a hazel-nut, having an apparently broad base, rough, and of a fresh red color. The second case is that of a Polish Jew, of about forty-five years of age, with perfect aphonia. There is a growth as large as a hazel-nut, and resembling a raspberry, seated upon the anterior attachment of the vocal chords, and accompanied by a secondary catarrh of the entire larynx. This formation covered entirely the anterior end of the right vocal chord, and a part of the left. In utterance the vocal chords passed behind this structure, lifted it, by giving it a leverage movement upon its attachment, and pressed it partially into the two ventricles of Morgagni. These might both prove to be epithelioma, according to the microscopic examination of the two cases extirpated by Lewin.

In respect to judging of the size of such tumors, I would call attention to the fact that the apparent size must vary considerably according to the position and the angle of attachment of the mirror. The last mentioned structure seemed, with a perpendicular position of the mirror, to fill up the entire glottis : and upon placing the mirror horizontally, there was a space of a finger's breadth between the structure and the posterior wall of the larynx. I would remark here that the two last named observations will probably also be again reported; for the first patient came to me after he had been for a time under the care of my colleague, Dr. Störk ; the second left me dissatisfied, because I stated that I had no powder, nor even a salve, with which to drive away the tumor, and, after the fashion of his countrymen, he has probably visited both Türck and Störk.*

---

* The translator begs leave to add one case, which came under his own observation, but unfortunately only at the post-mortem examination. G. R, a man forty-five years of age, " lost his voice " seventeen years before ; since that time he had never spoken aloud, and he was frequently troubled with catarrh. For some time before his death he could not speak above a whisper, and during the last few weeks he expectorated considerable matter of a very offensive character, streaked occasionally with blood. Suffocating paroxysms ensued, and the attending physician proposed to perform tracheotomy. It was not performed, however, and the patient died in one of these paroxysms. The examination revealed a morbid growth occupying the entire left side of the larynx, above the vocal chord, and somewhat below it; the individual parts were so entirely destroyed that it was impossible to recognise them. The arytenoid cartilage of the left side was immovable. The growth extended across the middle line anteriorly, dropping below the attachment of the vocal chords, and had begun to involve the right vocal chord in

As regards foreign bodies in the larynx, there is nothing to report laryngoscopically.*

Secondary Diseases of the Larynx Accompanying Acute Diseases.

Measles.

Stoffella has examined a number of such patients in Prof. Hebra's ward. The question was, whether the disease occurred in the larynx in the form of flecks, as upon the skin, or as a more evenly diffused redness and swelling—in short, whether there was a catarrh or an exanthema in the larynx.

He always found an evenly diffused redness of the mucous membrane of the larynx, and upon the vocal chords only a yellowish, or a reddish yellow coloring. But the patients were so sensitive to the irritation caused by the examination, that, after a few minutes even, an intense redness covered the vocal chords.† The redness extended into the trachea, but never beyond the fourth ring, and it was at the same time evenly diffused; only once was there found a dark red fleck about the size of a lentil upon the anterior wall of the trachea.

Variola.

The author has had the opportunity with Dr. J. Neumann, of examining a few patients with small-pox. The latter writes upon this point:

Variola in the air-passages is not particularly distinguished, either in its abundance or in its distribution, from that of the mucous membrane of the mouth. It is well known that in consequence of the variolous process, deep ulcerations occasionally occur on the above named spots. One case is recalled which

its anterior portion. The right arytenoid cartilage was movable. The translator regretted very much that he had not the opportunity of making a laryngoscopic examination, as a correct diagnosis was otherwise quite out of the question. He made a microscopic examination of portions of the growth, and found it to be an epithelioma. The specimen is in the cabinet of the R. I. Medical Society.

* Gibb reports two cases of foreign bodies discovered in the larynx by means of the laryngoscope. In one case a pin was removed with success. (Tr.)

† This agrees also with some observations which I have made.

terminated fatally in the *studium floritionis*, in which the mucous membrane of the pharynx and of the trachea was completely deprived of its epithelium and was covered with a layer of pus half a line in thickness, beneath which the reddish-brown mucous membrane, swollen for the most part to the thickness of a line, was eroded with the pits of variola. These alterations extended into the bronchi, as far down as the divisions of the third grade, whilst the mucous membrane of the finer branches appeared dark-red and swollen.

Scarlatina.

As yet there are no laryngoscopic observations of this disease.*

Erysipelas.

One might expect to find laryngeal manifestations, in some cases even of a severe nature, in connexion with erysipelas, and that too not merely of the face ; but this seems to be rarely the case, and only to take place in connexion with certain epidemics.

Since having my attention directed to this inquiry, I have not had much opportunity of examining patients with this disease ; among five patients whom I examined, four had erysipelas of the face; two were men ; one primary, one as intercurrent with a contused wound of the fingers : two were women, and in these it was primary. In four cases there was inflammatory redness and tumefaction on the epiglottis, and in the entrance to the larynx down to the vocal chords ; there was no oppression, no alteration of the voice. With the continued desquamation of the skin, the inflammatory appearances in the larynx also gradually vanished. In one of these cases a relapse of the erysipelas occurred, and so also of the catarrhal inflammation in the larynx. The fifth case was that of erysipelas in the lower limbs of a man who had an abscess on the leg, without any apparent alteration in the larynx. I propose to investigate this subject still farther.

* Gibb reports some cases. (TR.)

Lewin found in the larynx, at a post-mortem, tumefaction of the ary-epiglottic ligaments and of the posterior laryngeal wall, and intense redness of the vocal chords.

Lewin asserts that acute skin diseases may also give occasion to the development of neoplasms.

Typhus.

All the inflammatory diseases of the mucous membrane, of the sub-mucous tissue, and of the cartilages of the larynx, which we have considered in the preceding pages, may occur in the different stages of typhus; more frequently, however, in the exanthematic variety. The laryngeal diseases accompanying typhus have, according to their form, a very variable significance for the patient. Although in themselves possessing very little that is characteristic, their diagnosis still offers no particular difficulty, as may be easily imagined. In some epidemics severe diseases of the larynx occur more frequently. Laryngoscopic examinations of typhous patients exist in but very small number; I shall be pardoned therefore if, referring to what has already been said, I follow the representation of Rühle in that which succeeds.

Rühle observed acute catarrh several times, and in some cases erosions upon the vocal chords. Such often heal, indeed, without leaving any trace. So, too, croupous and diphtheritic exudations are found especially upon the posterior wall above the musculus transversus.

Twice Rühle found serous and sero-purulent infiltration of the sub-mucous tissue. The œdema of the entrance to the larynx, and of the vocal chords, which often renders tracheotomy necessary, is also, aside from catarrh, met with at the climax of the disease and joined with ulcerations, which may proceed either from the mucous membrane or from the perichondrium; it may also be collateral with parotitis. Sometimes, moreover, during the period of convalescence there may be a high degree of œdema succeeding the slightest exciting causes. The explanation of this is found in the constitution of the blood of these patients and in the feebleness of their circulation. Emmert has already called attention to this point.

Small abscesses were found a few times in the sub-mucous tissue, although one could not determine their relation to the typhus.

Perichondritis, the most severe affection of all, also occurs the most frequently in the period of convalescence. It may also be induced by the degeneration of diphtheritic ulcers upon the mucous membrane; it is frequently accompanied by œdema, and generally involves the necessity of laryngotomy. Its most frequent seat is upon the cricoid cartilage; it results in the formation of large cavities of pus with maceration and necrosis of the cartilages, which may, according to the location of the abscess, fall out by piecemeal.

Ulcerations occur at the height of the disease, even in the febrile stage in cases of great exhaustion, and render the prognosis more unfavorable. They proceed from diphtheritic inflammations (upon the posterior wall), or from the discharge of perichondrial abscesses, and also from the degeneration of typhous infiltrations of the mucous membrane on the posterior wall, which is very granular. A number of the laryngeal ulcerations in typhus are doubtless to be ascribed to decubitus, as they are found most frequently on points which are subject to pressure, as upon the lateral edges of the epiglottis, and at that point on its under surface, which, in swallowing, lies upon the apices of the arytenoid cartilages; also upon the internal side of the arytenoid cartilages, especially at the points of the vocal processes. The formation of decubital ulceration is still further favored by the tumefaction of the mucous membrane, by the diminished activity of the heart, and of course, also, by disease of the perichondrium.

What has here been said with regard to the existence of ulcerations in the larynx, is not only true of typhus but also of tuberculosis. Lewin draws attention, in a paper on this subject, to the fact that certain parts of the larynx are, by their anatomical nature and by their physiological function, particularly inclined to disease. Here are especially to be mentioned the very glandular mucous membrane upon the anterior surface of the arytenoids, and the point where the apices of the vocal processes are concealed in the vocal chords.

The more slight diseases of the larynx, and even the loss of the voice, may be readily overlooked in the case of typhus in consequence of the general condition of these patients. The

appearances of the various forms have been given in the preced
ing pages.

Rühle has seen good results follow in cases of catarrh and
typhous ulcerations, from cauterizations with solutions of nitrate
of silver ascending as high as a drachm to the ounce. Upon the
whole the treatment of the typhous diseases of the larynx has
nothing peculiar.

Secondary Diseases of the Larynx Accompanying Chronic Diseases.

Tuberculosis.

All the above-mentioned diseases of the larynx, from a simple
catarrh even to a neoplasm, may occur in connexion with tuberculosis; but the relation is very different from that which exists
e.g. in typhus. As a very frequent phenomenon we have to
mention anæmia of the larynx, which sometimes exists in connexion with ulcerations of certain spots; I might even ascribe a
prognostic value to anæmia of the larynx if there is no apparent
cause for it. Catarrhal inflammation is frequent, both as suba-
cute and chronic. The first appears as an inflammatory halo
around ulcers, or independently in connexion with tracheitis and
bronchitis, or as the forerunner of ulcers.

Türck described two such cases. Once a deep injected redness appeared
on both vocal chords, with subsequent formation of ulcers. Again considerable inflammatory tumefaction of one ventricular chord existed in a man
thirty-eight years old, in whom the most careful and repeated examination
of the lungs could detect no infiltration in their apices. Subsequently ulcers
were formed upon both vocal chords, and there was tuberculous infiltration
of the summit of the lungs.

In tuberculous patients the structures of the pharynx are
frequently so sensitive, that the examination becomes very difficult; in some cases entirely impossible. In some cases we may
still succeed by employing small mirrors, and by proceeding
slowly and cautiously; in others, by introducing the mirror
hastily, not allowing one's self to be terrified by a little straining effort in swallowing, and insisting upon quick inspirations.
I can only coincide with these remarks of Türck's.

Chronic catarrh often exists by itself a long time, either
without leading to the formation of ulcers or before their forma

tion. It often leads to ulceration of the follicles on those points where they exist in abundance; thus upon the posterior wall and upon the base of the epiglottis; also at and below the anterior attachment of the vocal chords. It may also give occasion to a callous constriction of the sub-mucous tissue, which manifests itself upon the epiglottis by thickening, upon the base of the larynx (the posterior wall) by the development of prominent swellings, upon the vocal chords and the small cartilages by thickening, with redness or with a pale color. Very frequently, then, some chance or other gives an impetus to the ulcerative process, and to its further unfavorable course.

We have here, therefore, again an important circumstance in the consideration of laryngeal diseases, with respect to tuberculosis: *i. e.* the existence of the frequently small, generally sharply bounded infiltrations of the mucous membrane, of a dull greyish color, upon which the ulceration commences in the form of points, which afterwards run together; thus the ulceration at the same time extends in depth. Upon the basis of the catarrhal mucous membrane in tuberculosis, the catarrhal ulcer frequently occurs, as well as those which are designated by Türck simple ulcers.

Besides ulcers proceeding from infiltration of the mucous membrane and of the follicles, decubital ulcers are found very frequently in tubercular cases. Here belong in part the ulcers upon the commissure of the arytenoid cartilages, in so far as this point is subjected to pulling or pressure by the movement of the cartilages; and particularly are to be included here the ulcers upon the apices of the vocal processes; in both cases the inflammatory tumefaction and sponginess of the mucous membrane of these points is expressly to be kept before the eye. Ulcerations must appear all the more readily upon the apices of the vocal processes, as according to Rheiner, the ossification of the arytenoid cartilages begins at the vocal process, in consequence of which its apex frequently runs out into a bony prong as fine as a needle. Rheiner also found here a small abscess between the mucous membrane and the cartilage. These ulcers extend in shallow furrows along the edge of the vocal chords, and this form is caused by the histologic character of the vocal chord, as its elastic fibres run in chord-like stripes parallel with the free edges. So also the edges of the vocal chords are found superficially indented, or filaments and small shreds are found hanging to them. Besides these, diphtheritic and croupous exudations are observed among tubercular patients when they are losing ground; so also aphthous ulcerations in the larynx, and upon the posterior wall of the trachea. Ulcerations, besides, being found upon those points mentioned, viz. the epiglottis, the vocal chords, their anterior attachment, and the anterior surface of the posterior wall, are also found upon the ary-epiglottic folds, the ventricular chords, and upon the posterior surface of the arytenoids looking towards the pharynx. By the simultaneous existence of such ulcers, there exists frequently among the tuberculous an uninterrupted girdle-like ulcerated surface, which surrounds the whole interior of the larynx, and can of course only be partially seen during life. The ulcerations upon the base of the epiglottis, and upon the anterior attachment of the vocal chords, generally extend themselves superficially; the former seldom produce perforation, and then only for a very small space, they do not commonly destroy the edges of the epiglottis. The surface of these ulcerations upon the epiglottis is seen only exceptionally; and that of those upon the anterior surface of the posterior wall is never seen; their edges only are seen; and they are often only to be recognised by the deposit, and by the projecting infiltrated points irregularly hanging over their edges. Laryngeal ulcers in tuberculosis are but exceptionally cured among us, yet each one of us has indeed seen such cases. Upon the basis of ulcers in tuberculosis there may arise small papillary growths of connective tissue.

Perichondritis, with its sequelæ, appears in tuberculosis always to be excited from without, by an extension of the ulceration in depth. It may lead to necrosis of the cartilage and to expulsion. We may sometimes recognise the loss of the arytenoid cartilages laryngoscopically from the sinking in of the posterior laryngeal wall, or from the deficient movement of the cartilages themselves. These phenomena are more manifest when they are unilateral; they may, however, be concealed by inflammatory tumefaction.

Türck makes the remark, that in patients who have been very much reduced by advanced tuberculosis of the lungs, paroxysms of dyspnœa are more easily borne; while laryngotomy is unavoidable in some cases in which the tuberculosis of the lungs cannot even be recognised.

Tuberculosis can certainly be recognised in some cases with the laryngoscope, before it can be demonstrated in the lungs, whereby the presumption of course is left open, that it exists just as well, but only cannot be demonstrated at that time in the lungs. What, then, are the laryngoscopic signs of tuberculosis? We reply, anæmia, infiltration, and ulcerations upon the points where there are numerous glands, *i. e.* on the basis of the epiglottis, and upon the anterior side of the arytenoids, in so far as typhus and syphilis can be excluded; also, with great probability, the flat ulcers upon the anterior attachment of the vocal chords, and the ulcers upon the points of the vocal processes, of which Lewin says: "No case has yet been presented to me, where the existence of tuberculosis could not have been afterwards verified if these ulcerations were actually present." It is scarcely necessary to add that corresponding to these existing alterations, more or less marked disturbances in the formation of the voice, in the functions of swallowing and respiration, pain, coughing, and expectoration must be present.

Three cases from my own experience I will mention on account of some peculiar circumstances.

A peasant, fifty-two years old, tolerably vigorous, in whom the apices of both lungs presented a slight infiltration, showed laryngoscopically, swelling and redness of the epiglottis, ulcerations upon both vocal processes, upon the right ventricular chord, and upon the anterior surface of the posterior wall. With such a larynx as this, the patient still blew a trumpet in the church on Sundays, *con amore.*

In the case of a house-porter, fifty-two years old, I was tempted, upon first introducing the mirror, to think of cancer. The greater part of the left ventricular chord, and of the ary-epiglottic fold, together with the coverings of the left arytenoid cartilage, presented a uniform greyish-yellow mass, highly fetid, evidently degenerating, and having a dirty appearance. Upon a closer examination I found a deposit upon the edge and upon the projection of the epiglottis, a degenerating infiltration about as large as a lentil upon the right ventricular chord, and upon the inner surface of the right ary-epiglottic fold. This and the examination of the lungs led to the correct diagnosis.

A short time since I examined a man about thirty years old. He had two years before a syphilitic ulcer, afterwards an eruption, and was treated anti-syphilitically with success. The examination of the summit of the lungs showed a somewhat shorter resonance and undecided respiratory murmur.* A more satisfactory examination could not be made, as I had to converse with the patient through the aid of a Dolmetsch. The laryngoscope showed anæmia of the laryngeal mucous membrane, with the exception of the reddened and swollen coatings of the arytenoid cartilages; upon the anterior surface of the latter there were three pointed, thick, tooth-like projections, between which a whitish-yellow deposit was visible; on the apex of the right vocal process there was an ulcer as large as a lentil, with thickened edges, and a pale and depressed base. The diagnosis was, most probably, tuberculosis.

Must I speak of the therapeutics of tuberculosis? Unfortunately, Madeira, Corfu, Venice, are accessible to a very small number! Of late, Kissingen seems to be in some repute. At all events, where there are but slight alterations, a change of residence may be made with some prospect of a favorable result. That suffocative paroxysms may sometimes demand an operation which shall make death more easy, is well understood.

Syphilis.

With this disease, also, there are frequent and very various diseases of the larynx combined, which, in their development, show much that is peculiar.

Besides a few cases by Czermak, Türck, Störk, and Lewin, we have before us a survey of syphilitic ulcers, mucous growths, and perichondritis, by Türck, and a still larger treatise by Gerhardt and Roth.

The existing data upon the frequency of certain diseases of the larynx in syphilitic patients, and the relation in point of time between their appearance and the continuance of the general disease, partially contradict each other and many of the

* " All those respiratory murmurs which offer no explanation of the condition of the parenchyma of the lungs, I name undecided respiratory murmurs," *unbestimmtes athmen.*—SKODA. (TR.)

general views upon so-called secondary syphilis; it is, however, still to be observed, that Gerhardt and Roth examined comparatively more of the earlier forms, others of the later; the former pursued their investigations in a syphilitic ward; and moreover, patients in whom the exanthemata or condyloma upon the anus and the external genitalia exist, are not generally sent to a laryngoscopist for the confirmation of the diagnosis.

Gerhardt and Roth say that when the appearances of constitutional syphilis upon the body pass beyond indolent tumefaction of the lingual glands, laryngeal affections may exist at any moment. These appear, either with or without affections of the pharynx, as acute catarrh, as partial œdema, or as condyloma, with a broad base, afterwards as isolated or extended ulcerations, to which the formation of irregular cicatrices is associated, and again as nodules, which quickly become purulent, or as gummy productions.

The localization of syphilis in the larynx is very common; for, of fifty-six syphilitic persons whom Gerhardt and Roth examined, eighteen, i. e. one-third, were thus affected. Whether women or men are the more frequently affected in this way is not yet determined; among children these diseases seem to be very rare, yet there are a few such observations. One-half of the syphilitic diseases in the larynx observed by Gerhardt and Roth were in the third decennium of life. Of forty-four persons with the earlier forms of syphilis, eleven (or one-fourth) were affected by the disease in the larynx; and of twelve with the later forms, seven (or more than one-half).

As more immediate causes, we may trace the earlier forms to over-exertion of the voice and to colds, while for the later forms these causes and insufficient mercurial treatment may be blamed. Severe diseases of the larynx may arise and may be cured by mercury or the iodide of potassium, many years after the primary diseases can be recognised, or when from the examination, or from the anamnesis, there are no grounds for arriving at the diagnosis of the syphilitic nature of the affection. In such cases we must not forget, on the one hand, that the first appearances of the syphilis (induration and exanthema) may be easily over-looked; and on the other, that a person may be inoculated with syphilis without any syphilitic ulcer being recognisable, although the inoculation of syphilis most frequently occurs by means of the ulcer (Reder, Pathology and Therapeutics of Venereal Diseases. Vienna, 1863). Finally, cases of intentional concealment are to be borne in mind.

As the most frequent type of the earlier laryngeal diseases in syphilitic patients, Gerhardt and Roth recognise the papillary formations, which are uneven, whitish, flat or pointed projections of various size, on various parts of the larynx, and in various extent. ·Czermak, Türck, Gilewsky, et al., have also observed some such growths of cellular tissue, resembling condyloma ; so also excrescences, an uneven character in the vocal chords, tumors with broad bases, and *plaques muqueuses.* Gerhardt and Roth declare that these formations are certainly syphilitic, because they have never been observed in entirely healthy people, nor in those affected with other types of laryngeal diseases. For what formerly were described as growths of cellular tissue, in connexion with tuberculosis, are papillary growths in the vicinity or upon the basis of a long-existing ulcer ; whilst in the cases of Gerhardt and Roth ulceration was never found existing at the same time. They consider these formations as broad condyloma, because they show great similarity to the latter, especially in the whitish thickening and sponginess of the epithelium, just as in the condyloma of the mucous membranes. Moreover, in all these cases, with one exception, there were found very extensive condyloma elsewhere, and especially in the buccal and pharyngeal cavities. These condyloma were seated the most frequently upon the vocal chords and upon the folds between the arytenoid cartilages, as well as upon those places which by frequent friction were mechanically irritated. Twice they found these condyloma upon the posterior wall of the pharynx, opposite the point of the epiglottis ; they could not therefore be recognised with the mirror. Twice colds were recognised as the exciting cause.

In respect to the time of their appearance (from the sixth to the tenth week), and to their symptomatic significance, Gerhardt and Roth place the condyloma in the same rank with catarrh, which has only been observed by them twice, and offers nothing of special importance. These statements upon the frequency of catarrhs, and of the broad condyloma in the larynx of the syphilitic, are entirely contrary to views hitherto held. Of Czermak's cases, only one, the fifteenth, is to be classed decidedly among the broad condyloma. Türck asserts that he has seen only one such

decided case ; Lewin cannot confirm the frequency alleged by
Gerhardt and Roth. I have myself had occasion lately to
observe, in the hospital, twelve syphilitic patients, all of whom
had broad condyloma upon the sexual parts, upon the anus, etc.,
and among those I saw four cases of catarrh of the entrance of
the larynx ; once also a finely-streaked injection of the vocal
chords, but no growths of the mucous membrane. The recon-
ciliation of these opposing statements must be left to further
investigation, unless we would make the assumption that Ger-
dardt and Roth regard as broad condyloma, structures which we
here would not recognise as such.

Ulcerations in the larynx proceed from various causes. The
superficial ulcerations proceed in most cases from broad condyloma (Reder). Deeper-seated ulcerations proceed from a suppu-
rating perichondritis, or from the degeneration of nodular deposits. The superficial ulcerations, according to Gerhardt and Roth,
appear to be seldom, and to occur later than had been hitherto
supposed ; they are not to be considered therefore in the same
light as the ulcerations of the pharynx. These ulcerations occur
on many places, but very frequently upon the epiglottis and the
vocal chords, with or without a manifest halo of inflammation ;
as a rule, some disease of the pharynx seems to have preceded
it, but it is by no means always simultaneous. Just so we sometimes find the epiglottis entirely free, while there are ulcerations
in the larynx, and simultaneous cicatrices or ulcerations in the
pharynx, so that the assertion that the laryngeal ulcerations of
syphilitic patients proceed from an extension from the pharynx,
does not apply to all cases.

Among their cases Gerhardt and Roth found nodular deposits on and in the larynx three times, once on the posterior wall
of the pharynx, and once also upon the epiglottis. Lewin found
such a tumor resembling a mulberry, seated in the right glosso-
epiglottic ventricle, about 11 millimètres (0.42 inches) broad and
9 millimètres (0.35 inches) high, which produced asthmatic
trouble from pressure upon the epiglottis, and became smaller
by the application of chromic acid. Under this head belong
some of Czermak's cases.

The laryngeal ulcers in syphilitic patients have very little
that is characteristic, save that on the epiglottis they especially
attack its edges, and, as a rule, produce a greater loss of sub

stance, a peculiarity which they share with lupous and cancerous ulcerations, but which occurs only exceptionally in tubercular complaints (Türck).   Upon the cicatrized edges of old ulcers of the epiglottis, the edge of the cartilage may often be seen shimmering through, as a white stripe.

Still further, there is occasionally found with syphilis, sclerosis of the mucous membrane and of the sub-mucous cellular tissue, and the formation of papillary excrescences upon the base of ulcers I think, as well as accompanying other superficial ulceration.   The sclerosed mucous membrane may be destroyed to a great extent, or the ulcerative process may extend in depth, giving rise to perichondritis, with purulent formation, and at last a discharge inwardly may ensue, together with necrosis and the loss of pieces of cartilage.

This generally destructive form of laryngeal syphilis is the best known.   None of the cases on record occurred before the twenty-first year.   The general character of these cases is the stenosis, which often necessitates operative interference, and may even lead to perfect closure of the laryngeal cavity.   Their essential characteristics are: the longer existence of the syphilis; the repeatedly intermittent hoarseness, dysphagia, cough, dyspnœa; finally, loss of voice, pain in the throat, a mucopurulent, bloody expectoration, emaciation, and danger of suffocation.   If life even is saved by general treatment, or by laryngotomy, permanent disability generally still remains.

I performed tracheotomy in consequence of dyspnœa, with good success, in the case of a woman thirty years old, in whom the traces of syphilis in the pharynx and larynx were unnoticeable.   The entire edge of the velum and its posterior arch had become adherent to the posterior wall, and they are still so, being pierced in the centre by a hole as large as a copper; the right edge of the epiglottis and its apex is in like manner attached; the left edge is wanting, and so also the left ary-epiglottic fold. Through this space a little piece of the left ventricular and vocal chords are seen; the latter is movable, and white; the point of the constriction is not to be seen through this opening, which is as large as a lead pencil.   At the time of the operation the constriction occupied the region of the cricoid cartilage, for after opening the conoid ligament (the middle crico-thyroid, Tr.) the canula could not possibly be inserted; the finger, however, easily

felt the vocal chords above, but it could not be introduced downwards. The patient had in the meantime become apparently
dead. I passed a blunt bistoury through the constricted place,
separated it as far down as the first tracheal ring, and easily
inserted the canula : then by the use of artificial respiration, we
saw the patient revive in the course of a half hour. Slight
inunction was made. At present the larynx is narrowed in the
sub-glottic region ; the patient is obliged to wear the canula, and
is entirely voiceless. The entire course of the disease may be
greatly altered by the intercurrence of acute affections, of perichondritis or œdema.

Laryngeal ulcers in syphilitic patients, generally leave behind
them loss of substance (see above), and radiating chord-like
cicatrices which project, causing constriction, together with lasting functional disturbance. Cicatrices upon the vocal chords or
upon the arytenoid cartilages produce incurable alterations of
the voice. These may escape notice in examination from various
reasons, among others in consequence of an unfavorable position
of the thickened epiglottis, or from its being made fast by cicatrices lying below, in front of the arytenoid cartilages. I only
need mention that just as in the larynx, so also in the naso-
pharyngeal space and in the nasal cavities, the syphilitic diseases
of the mucous membrane and of the bones occur, and in this
relation I would refer merely to Figs. 4, 5, and 6, of Plate II.
In a syphilitic patient who had condyloma upon the anus, I
have once seen catarrh of the pharynx, and of the entrance to
the larynx, and ulceration of the tonsils, with a similar growth
upon the nasal mucous membrane as is depicted in Fig. 3,
Plate I.

The prognosis of the syphilitic diseases of the larynx
depends upon the existing alterations. The treatment is a general
anti-syphilitic one, with mercurials or with preparations of
iodium, and it often gives very satisfactory results.

I had sent to me by my highly esteemed instructor, Prof.
von Dumreicher, a patient who was suffering from considerable
dyspnœa and hoarseness, and who bore upon him the marks of
syphilis. The examination showed that the epiglottis, the right
ary-epiglottic fold, and the right ventricular chord, were manifestly reddened and swollen almost to the thickness of the thumb ;
the right vocal chord was but slightly movable, with its edge

alone projecting, its sub-glottic region in like manner reddened, swollen, and projecting. These swollen points of the mucous membrane did not appear at all œdematous, but they gave the impression of a dense infiltration. There was no ulceration. The diagnosis of syphilis was from various circumstances made with accuracy : the dyspnœa had of late rapidly increased, so that each moment a suffocative paroxysm was to be apprehended. If, therefore, tracheotomy was to be avoided, immediate and vigorous treatment must be at once resorted to. I put the patient to bed and commenced an active inunction cure. The dyspnœa increased, however, until the second succeeding morning, and then quickly diminished. On the eleventh day there was a well marked stomatitis with salivation, but the dyspnœa had entirely vanished. The stomatitis required nearly three weeks for its cure under the local application of a solution of iodium and glycerine; after a time the patient improved, and even now, after the lapse of two and a half years, he is very well: the swelling in the larynx has disappeared, the right ventricular chord is still a little prominent, the right vocal chord is but little movable, and hence the disturbance of his voice is lasting.

A short time since a colleague brought me a patient who had been hoarse for some time, and had difficulty in swallowing. There was dulness in the apices of the lungs. Following medical advice, the patient had spent the summer in Gleichenberg, but without avail. He had recently been to Türck, and was now brought to me, because Türck's diagnosis had not proved satisfactory, but at the same time I was told that Türck's opinion would not be stated to me, until mine had been heard. I found upon the left half of the much reddened and thickened epiglottis, an extensive loss of substance, together with catarrh of the larynx, and an ulcer with an uneven base upon the right arytenoid cartilage. I said that according to the present appearances I should consider it as syphilis, and was told that Türck had also made the same diagnosis. By this similar decision the patient and his physician were convinced, and determined to commence corresponding treatment. This case affords proof of the certainty which laryngoscopic diagnosis has already reached.

Chlorosis.

I think I have proved by two cases cited under the head of symptomatic chronic catarrh, that secondary diseases of the larynx may occur in connexion with chlorosis.

Scrofulosis.

Scrofulosis gives occasion in some cases to diseases of the larynx. Hitherto there have only been observed sclerosis and increase of volume of the mucous membrane, and of the sub-mucous cellular tissue; these may produce obstinate constrictions, but they may also without doubt proceed to ulceration.

Such a case is the 1st of Czermak's (see examination of the larynx from below, page 97).

A case belonging to this category I described in my first essay upon the laryngoscope, and as it was the first pathological case observed with the mirror, it may properly be introduced here.

A healthy looking girl of fourteen years of age, at the ambulatorium of Prof. Dummreicher's clinic, having ulcerations upon the posterior wall of the pharynx and upon the soft palate, was treated with a weak solution of iodium and glycerine. A cure ensued, with an adhesion of the soft palate to the posterior wall. Ten weeks afterwards she came again with a hard, uneven, yellowish-red nodule of two inches in length, placed longitudinally upon the dorsum of the tongue, and covered over with enlarged vessels. In the course of a fortnight this changed into a large ulcer, with hard, thick, ragged edges, which burrowed deeply into the substance of the tongue, while in the anterior part, a new nodule was developing itself. A fissure of the ulcer extended three-fourths of an inch behind the foramen coecum, and there terminated in the shape of a star. Suspicion of syphilis. Negative result both of the examination and of mercurial treatment. Painting with a solution of iodium and glycerine twice a day, three grains to the ounce, by the aid of the mirror: internally a scruple of iodide of potassium daily. Cure after a few weeks.

For simple lupus the development and the cure was too speedy. In such infiltrations, the diagnosis, especially between syphilis, lupus, and scrofulosis, is often very difficult. With regard to diseases of the larynx in connexion with other chronic diseases, there are no laryngoscopic observations.

11

Symptoms of Diseases of the Larynx.

The most important symptoms in diseases of the larynx, are alteration of the voice, dyspnœa and dysphagia, coughing and pain. It is generally known that but few diseases of the larynx exist without producing

Alteration of the Voice.

We have in another place (page 77–79) considered what speech is, and set forth as the components of what we call speech, tones, murmurs, and a peculiar resonance, which gives each man a voice peculiar to himself, and enables us to distinguish friend from foe even by their speech. It is evident from what has been stated that, in the consideration of the voice, there are many and difficult questions arising, which still await a satisfactory answer. Loss of voice is distinguished from the tones of speech by the absence of a musical tone. Hoarseness occupies a middle point between the two. The quality of the tone of the voice depends upon the relations as to size and space of the organs producing the voice on the one hand, and of the windpipe and mouth-piece on the other; that is to say, upon all those parts which are brought into vibration when a tone is produced, as the chest, the trachea, the larynx as a whole, the structures of the mouth, nose, and pharynx, together with the contained air.

The study of the variations of tone in the voice is very arduous. Lewin has made a beginning. He says the pathetic hollow tone of the voice is caused by depression of the epiglottis by a certain degree of effort. Hence there is found among preachers, who speak the most frequently in this voice, a manifest depression and bowl-shaped curvature of the epiglottis, and combined with this a material thickening and bluish-redness of the ary-epiglottic folds; still these are by no means the only signs of hoarseness among preachers. Farther, the voice of command of officers has a shrill, trumpet-like character. This peculiarity of the voice arises from the fact that the cartilage of the glottis opens to the powerfully expelled current of air; the air rushing out, rubs over the mucous membrane of the posterior wall of the larynx ; and when once catarrh exists, it may produce permanent thickening and swelling upon those points, so that then a small mucous fold presses into the posterior extremity of the glottis, and thus a triangular chink remains open, a condition which is not unfrequently met with among officers who boast of a fine voice for command.

The tone of the voice does not further concern us, nor do the alterations of murmur, the so-called defects in the speech. We have to do with the alteration or with the absence of the musical properties of speech. Every increase or diminution in the requisites of the voice (see page 77), must lead to alteration in the voice.

I will commence with a *diminution of the current of air.* This is insufficient for the production of a musical tone, either when too little air reaches the vocal chords, or when it does not move with sufficient rapidity. If a person has a physiologically perfect glottis, but beneath it an opening in the air-passage, as after laryngotomy, he will not be able to produce any tone, because the air escapes through the opening and an insufficient quantity reaches the glottis. It would be the same with those persons who wear a tube after the artificial opening of the air-passage ; hence for the purposes of speech, the closure of the tube must be provided for. A diminution of the current of air also takes place, when either by a neoplasm in the interior, or by the pressure of a tumor from without, the diameter of the air-passage is materially diminished. But upon the one hand other symptoms will here demand more of the attention of the practitioner, and on the other the patient may, in various ways, conceal this deficiency more or less by increasing the action of the diaphragm and of those muscles which contract the chest, or by speaking in short sentences. A diminution of the rapidity of the current of air—of the pressure of the column upon the vocal chords—will also affect the production of tones. Those persons in whom respiration is attended with pain or is performed with difficulty, will never speak loud, even though the larynx may be physiologically perfect; if the movement of the diaphragm is hindered the necessary pressure may be yet brought to bear, but only for a short time; hence, if they would speak loud, they can only do so in short sentences.

We will next consider hoarseness and aphonia as it proceeds from *an insufficient closure of the glottis.* The closure of the glottis occurs from behind forwards, first, by an approximation of

the arytenoid cartilages from a contraction of the posterior arytenoid and of the superior fibres of the posterior crico-arytenoid
muscles; and second, by the two arytenoid cartilages being prevented from rolling outwards; in other words, by their rolling
inwards until their inner surfaces and the vocal processes touch
completely; this is accomplished by the action of the lateral
crico-arytenoid muscle. We have a whole category of alterations of the voice from an insufficient closure of the glottis in
consequence of alterations in the motor apparatus itself, from
atrophy of the muscles in chronic catarrh, from paralysis in
consequence of alterations in the brain or nerves, from pressure,
&c.

Imagine the lateral crico-arytenoid muscle of one side rendered incapable of action, and an attempt made to close the glottis for the purpose of giving utterance to a tone. One vocal
chord draws near to the middle line, while the arytenoid cartilage elevates itself; the arytenoid cartilage of the diseased
side remains more or less in the condition of being rolled outwards, as long as the opposition of the thyro-arytenoid and
posterior crico-arytenoid muscles to the posterior arytenoid continues. The vocal process of the diseased side remains directed
outwards. If, as is commonly the case, the healthy vocal chord
overlaps the middle line, there is a triangular space left open
which is too large to allow of the production of musical vibration by the expelled current of air.

Suppose that the posterior crico-arytenoid of one side was
inactive, with or without the posterior arytenoid; then the arytenoid cartilage of the diseased side remains rolled inwards, and
in the first case, at least in the middle line, approximated to the
other; in the second case it is also shoved sidewards. The vocal
chord stands either at or beyond the middle line; the apex of the
vocal process projects beyond it; if a closure of the glottis
should ensue, there would be so large a space left open between
the arytenoids standing apart from each other and the approximated apices of their vocal processes, that the current of air in
passing outwards could not produce a musical tone.

These conditions are generally found only upon one side, and
are easy to recognise. Sometimes they are found with tracheitis
or tuberculosis, when the connexion cannot be adequately explained. In some cases pressure, fear, &c., must be assigned as

the cause. There are cases recorded by trustworthy observers, in which the voice was restored in the same inexplicable way as it vanished.

Lewin once found a luxation of an arytenoid cartilage. By means of a catheter-shaped instrument the replacement was readily effected with immediate restoration of the voice. Stoffella once observed both the cartilages of Santorini grown together.

In the lower grades of hoarseness, frequently nothing farther is discovered than that the movements of the vocal chords occur more slowly and with less certainty. We might suppose in such cases that a portion of the expelled air has escaped before the requisite closure of the glottis occurred, and thus the hoarseness might be explained. Hitherto this condition has been termed " gaping of the glottis."

The closure of the glottis may be mechanically hindered by tumors of the vocal chords or of their vicinity, which, when they are pediculated, fall between the vocal chords just as a sound is made; by granulations and infiltrations upon the vocal chords, or upon and between those surfaces of the arytenoid cartilages which are turned towards each other (the ulcus elevatum of Stoffella); by tumefaction of the mucous membrane of the posterior wall, placing itself between the arytenoid cartilages (Gerhardt, Lewin); and by swelling of the ary-epiglottic folds and ventricular chords, whereby the movement of the arytenoid cartilages is obstructed, or the vocal chords pushed one side. There is not an elevation of the tone from vibrations formed by a shortening of the vibrating portion of the vocal chords, but there is hoarseness or loss of voice from the obstructed closure.

Czermak makes the valuable remark that small polypi upon the vocal chords often affect the voice more than large ones, inasmuch as the former frequently become wedged in the glottis; but when they grow larger a pedicle is formed, and they may be thrown upwards out of the glottis by the current of air, thus allowing the vocal chords to approximate each other more closely.

Traube observed a case of aneurism of the arch of the aorta united with hoarseness. He found the epiglottis injected, its gutter-like curvature more pronounced and pushed towards the left, the ventricular chords and aryte-

ncid cartilages reddened, the vocal chords normal; the left remaining perfectly still when a sound was made, while the right promptly moved towards
the middle line ; the arytenoid cartilages were similarly affected, and thus a
chink remained between the two vocal chords. The voice had also become
higher.   Trouble in swallowing.   In this case the left recurrent nerve was
paralysed by laceration.

I have lately, by the kindness of Prof. Duchek, seen a similar
case in the Joseph Academy, proceeding also from an aneurism of
the arch of the aorta.   The left vocal chord stood firmly in a
middle position, and the left cartilage of Santorini was lower
down than the right; or translated physiologically, the left arytenoid cartilage stood with its apex directed more backwards than
the right, and the action of the muscles, which close the glottis,
was suspended.   A thorough examination was prevented by the
great sensitiveness of the patient.

The glottis may, moreover, stand more or less obliquely, and
this may depend upon an oblique position of the mirror, upon
dislocation of the thyroid cartilage, or upon scoliosis of the
larynx, originating in itself or proceeding from tumors of the
neck (Gerhardt), and indeed without alteration in the voice.

I have mentioned that the division of the nerves in the
larynx is by no means clear.   Perhaps after more extensive
pathological observations some new light will be thrown upon
this point.

Small lumps of mucus upon the vocal chords act as dam-
pers when they become wedged into the glottis, and prevent its
closure (Gilewsky).   As a fact, it frequently occurs that while
speaking hoarseness suddenly ensues, and vanishes of itself after
a slight clearing of the throat or after a little rest.

Ulcerations, abrasions, and infiltrations, which must necessarily be pressed or torn by the closure of the glottis, may induce
the patient not to close it forcibly, but rather to speak without a
full tone.   Finally we have to mention loss of substance in the
vocal chords, and cicatrizations in so far as they cause a drawing-
in of the vocal chords, or hinder their motion and that of
the arytenoid cartilages.   A deficient closure of the glottis
offers an obstacle to expectoration.   Türck speaks of a phonetic
and a respiratory paralysis of the glottis.

The third cause of hoarseness and loss of voice lies in the
diminished tension of the vocal chords.   We have above stated

that in otherwise like circumstances the tone becomes deeper and finally impure by a diminished tension of the vocal chords. This may within certain limits 'be equalized by an increased force given to the expelled air; but there is a degree of diminution in the tension of the vocal chords, in which this equalization no longer takes place. It is only a little while since we have been able to distinguish the differences in tension of the vocal chords, and in fact it presents great difficulty. The difference between a closed and a tense glottis may be very well demonstrated with the requisite practice. When upon fixation of the point of attachment of the vocal muscle, its fibres contract, the elastic fibres and the mucous membrane are elevated and made tense, while the vocal chords are rendered firm and projecting. What might be said of passive longitudinal tension, corresponds so closely with the discussion of the action of the muscles in opening and closing the glottis, that we may here omit its consideration. The diminution of the active and the normal tension may be best considered under one head. The active tension, as regards innervation, may be diminished or increased by alterations in the corresponding nerves, as e. g. by enlargement of the thyroid gland, &c.

A simple inflammatory affection of the mucous membrane of the vocal chords may produce an effect upon their tension by causing sponginess and a greater degree of moisture. If we remember that beneath the extremely delicate mucous membrane there lies a very fine and firmly adherent layer of elastic tissue, which, being in its turn firmly adherent, covers directly the vocal muscle, we can infer that an inflammatory disease of the mucous membrane of the larynx cannot easily exist without being followed by an exudation into the layers beneath, as e. g. in catarrh, by a serous infiltration of the membrane and muscle of the vocal chord; similar results under similar circumstances we recognise in like manner in other parts of the body. Disturbance of the activity of the affected muscle is therefore a necessary consequence. It is however evident that in other forms of inflammation the indicated condition must occur in a still greater degree, even although the vocal chords themselves are not precisely the seat of the inflammation. Also in inflammation of the perichondrium or of the ossified cartilage, œdema of the vocal chords themselves may ensue. Similar anomalies in the formation of the voice occur in connexion with inflammatory diseases of the trachea.

Farther, it is easily seen that in the chronic forms of inflammation the disturbance in the play of the muscle must be lasting, for there ensues either organization of the exuded matter, *i. e.* thickening, or else absorption, and then atrophy of the vocal chords. These alterations remain behind in a greater or less degree, even after the cessation of a lingering catarrh; in the first case the vocal chords seem thickened at the expense of their breadth; in atrophy they are more or less reddened or discolored, and lie in longitudinal folds.

It is of course understood that in many, and even in most cases, the anomaly in the formation of tone is not exclusively to be traced to any one of the above mentioned causes, but that, on the contrary, two, or even all three, may work together, and thus the consideration of the subject becomes still more difficult. There will also be cases in which one cannot prove the existence of any one of the causes referred to, although they must be present.

It may be asserted, however, that as the inaction, or a diminution in the force, of these factors of the voice leads to its modification, to hoarseness, or to aphonia, so an excessive degree of activity must lead to a pathological alteration in the voice.

An excess of force in the air expelled will scarcely occur as a disease. When the glottis cannot be opened to a sufficient degree, a screeching voice will of course follow under otherwise like circumstances; just so a cramp or a contraction of the muscle of the vocal chord, which produces the longitudinal tension, will necessarily influence the voice unfavorably (see above); but in these cases other symptoms engross the attention of the physician. It must not be overlooked that within certain limits a compensation takes place among the several agents in the formation of the voice.

As one ground for an imperfect voice, or for entire absence of it, there might be adduced the congenital absence of the vocal organs. Although these cases seem to be seldom, yet they may occur.

Gibb reports in the *Medical Times* two such cases occurring at the West London Hospital, in persons born deaf and dumb. In one of these cases the epiglottis only existed as a small stump, the right ary-epiglottic fold formed a thick bolster, the left arytenoid cartilage seemed to be wanting; in the other case the entrance to the larynx was normal, but in both cases the vocal chords were wanting.

I confess that these observations are *à priori* suspicious.* We have indicated above the circumstances under which pain and dysphagia occur; also, what we had to observe upon coughing and expectoration.

Dyspnœa.

In the preceding sections we have had occasion to show how almost every disease of the larynx may produce dyspnœa by constriction of the air-passages.

Türck, in his review of this subject, arranges these constrictions in three classes, according to their seat.

(a.) Above the glottis. Tumors of the pharynx. Here a case of cancer is adduced. An uneven nodule, as large as a nut, was seated upon the posterior and left wall of the pharynx, and thrust itself beneath the epiglottis. In a second case, probably also cancer, the swelling was seated behind the arytenoid cartilages. Tumors at the entrance to the larynx (see above), and œdema of the ary-epiglottic folds.

(b.) Constriction of the glottis, from simple acute and chronic catarrh, from syphilitic inflammation, croup, inflammation and œdema of the vocal chords, with ulcers or with perichondritis, from abscess in the vocal chord itself, from cicatrices after injuries and after ulcerations, from neoplasms, and from abnormities in muscular activity.

(c.) Below the glottis, from an annular constriction without any more definite character. In one case which had lasted five months in a boy fourteen years old, blood-letting brought relief. The larynx was narrowed down to the thickness of a quill. A second case was in a young woman. The constriction was less marked, and was produced by a deposit of mucus and epithelium. Constriction at this point may also arise from perichondritis, from growths of the mucous membrane of the larynx after tracheotomy (observed by Türck and myself in typhous patients), or from tumors in the larynx and the trachea.

Türck observed a very peculiar case of constriction of the trachea with deficient respiration, in a young man. The larynx was normal, the trachea reddened, and at a very slight depth narrowed down to a very small chink

* The translator recently had an opportunity of examining a deaf mute. He found the left vocal chord entirely wanting, and through the space thus left an extensive view of the trachea was to be had. The right vocal chord performed the normal movements of respiration.

directed from forwards backwards, which was broader in expiration than in inspiration. If the patient uttered a shrill tone, the edges of this chink vibrated, while the vocal chords remained perfectly quiet, and stood far apart from each other.

Cases in which laryngo-tracheotomy was performed in consequence of dyspnœa, and which were observed laryngoscopically before and after the operation, are reported by Balassa and Ulrich.

The success of tracheo-laryngotomy is very often in so far imperfect, as great difficulties attend the removal of the canula after the expiration of the disease. If those persons who are obliged to wear a canula, would return to the general concerns of every-day life, there must be some way provided for the closure of the canula for the purpose of speaking, without calling the hand in aid. I will not here go into the numerous attempts which have been made in this direction by inventive physicians and mechanics, the various single and double valves used, etc., but I would make mention of the latest form of canula with a bullet-valve made by our valiant Leiter. It is a double canula of hardened rubber; upon the inner tube a little cup is fixed by friction, which is open on the one side towards the tube, and on the other forwards and outwards. In this cup a ball, also of hardened rubber, plays, following the current of the air; thus, by inspiration, it slips inwards, by expiration it is thrown against the anterior opening of the cup with a very slight rattling sound.

A patient who for a long time has in turn worn various canulas, has used this last one for more than four months with perfect satisfaction, and says that the air which passes through it is warmer than that through metallic canulas; this, in general, is more agreeable, and only when the temperature is very high, is he obliged to have recourse to other canulas. Moreover, the gutta-percha canula has proved itself to be more lasting than those of German silver.

Negative Results.

In the foregoing pages we have had frequent occasion to refer to the positive results of laryngoscopic examinations. Under some circumstances, however, negative results are no less important, whether it be because they allow us to relieve anxious fears which the patient may have had for long years, or because they give occasion to still farther investigations of the causes of existing disturbance, and thus frequently lead directly to interesting studies.

More than three years since I was asked to examine a patient at the hospital in the Wieden, who was suffering from dyspnœa. As I arrived all the preparations had been made for the operation of laryngotomy, as the attacks of suffocation were at times so violent, that at any moment it was thought the operation would have to be performed.

A careful examination of the patient had discovered no cause for the dyspnœa. The mirror showed that the larynx and the trachea, down to a certain depth, were entirely normal. After it was established that the cause of the dyspnœa must be seated very low down, we again made a thorough examination of the patient, and in consequence of a difference in the pulse of the two radial arteries, I advanced the opinion that there might be an aneurism of the arch of the aorta. The dyspnœa once more diminished, and as afterwards it again commenced, we could discern the aneurism of the aorta, from which the patient afterwards died: tracheotomy was not performed. Not only would the operation have been useless, but it might also have been directly injurious.

In a second case which was sent to me, I made the same probable diagnosis, but I can make no report on the subsequent progress of the case. It is well known that there are aneurisms of the aorta, the existence of which cannot for a long time be proved.

A short time since a man was sent to me, who was about forty-two years old, well built, and who complained of considerable difficulty of breathing, which increased very much by movement and by mental excitement. It had existed five months, having appeared gradually; then, according to the report of the case, the patient was taken sick with pleuritis, and afterwards the dyspnœa was more violent. During the summer the patient took good care of himself, and drank the waters of some spring or other. During this time his condition improved somewhat, still the difficulty was quite great, and, as an increase was to be feared upon the approach of cold weather, the patient came to Vienna. The anamnesis elicited, in addition to the above only, that he had suffered from a venereal ulcer three years before, which was cured by local treatment. There were never any signs of syphilis present.

The examination of the larynx and of the upper part of the trachea, which was quite difficult, revealed no disease, and furnished no explanation of the growing respiratory murmur, the cyanosis of the face, &c. The external examination showed that the anterior region of the neck and the upper part of the breast were irritated by tartar-emetic ointment, but nothing farther. In like manner percussion revealed nothing. On auscultation there was heard over the entire right lung, in front and behind, a groaning rough murmur, which became more marked towards the division of the trachea, and was very faintly heard over the left lung, the less distinctly the farther removed from the division of the trachea. There was therefore clearly a stenosis of the right bronchus.

It would have been very desirable to look into the trachea, down to its division, and into the commencement of the right bronchus; but I could not succeed in doing so, even after two sittings of an hour each, in spite of all my pains. I do not allow myself to make any supposition upon the kind and nature of this constriction. Latterly the condition of the patient has been greatly improved by an expectant treatment.

Czermak describes a similar case (No. 12); he also did not succeed in seeing the division of the trachea.

A boy fifteen years old went home one night from the music school in the city to Gumpendorf, and on the way, without any apparent cause, lost the power of speech altogether. The family were greatly excited. The boy was brought to us at the Surgical Ward of the Gumpendorf Hospital. The mother stated that the boy had at times spasmodic twitchings in the arms and legs, that he learnt with difficulty, and sometimes seemed to be quite lost. We found the face greatly reddened, the movements of the heart very violent, a systolic murmur, and farther nothing abnormal. The movements of the lips and tongue were good in all directions; the laryngoscopical examination showed the larynx to be normal, and the vocal chords readily movable. Although the entire history was in itself peculiar, this last circumstance adduced almost made a certainty of the suspicion of deception. As we observed that the boy was greatly perplexed by the minute examination, and grew red at many of the cross-questions, an attempt was made to draw out a confession. I told the would-be sick boy, sharply and decidedly, that he was a liar, and after a promise of secrecy, the suddenly-restored patient told me with tears that he did not wish to learn to play the violin, that he had been beaten at the school for having learnt nothing, and he hoped by this tragi-comedy to scape musical instruction for the future.

We could scarcely count the number of times that we have found catarrh of the pharynx when patients have feared that they were suffering from laryngeal phthisis.

Addenda.

We have already alluded to the importance of a careful examination of the cavities of the pharynx and the mouth. The external examination and touching of the larynx should not be neglected, as it affords very useful information as to the form of the parts, pain, &c. Auscultation and percussion are also called in aid for the diagnosis of diseases of the larynx. The critical labors in this department have lost much of their worth since the introduction of the laryngoscope.

In respect to the application of remedies in powder to the air-passages, I would add that I have lately made some experiments with pulverized charcoal. At first I breathed in through a quill a considerable amount, by a quick expulsion of my breath. I carried some of it, to be sure, into the larynx and the trachea, but the greater part remained lying in the pharynx. It is also not clearly established that every patient will bear the inhalation equally well.

If the charcoal is blown in with the tube above described, one can see the interior of the larynx, and of the trachea far down, dusted over with the powder, naturally thicker in the upper parts than farther downwards. The vocal chords were never found equally coated with the powder, but it lay at the attachment of the epiglottis and upon the anterior end of the glottis, the points where the secretions also accumulate. The subsequent examination with the mirror could never take place immediately afterwards, but we were obliged to wait for two or three paroxysms of coughing, which at the same time produced hyperæmia of the larynx and of the velum. The reaction, however, is limited to this.

TRANSLATOR'S APPENDIX.

THE following cases reported by Dr. Semeleder in the Vienna Medizinal Halle of January, 1864, were published in the *American Medical Times* in May by the translator. The novelty of these two cases, the light they throw upon physiology, the difficulty attending the operations, and the comparatively successful results obtained, afford, I think, a sufficient reason for making them available to the majority of American practitioners. The few passages which have been omitted possess a mere local interest.

Two Cases of Extirpation of Polypi in the Larynx.

Laryngoscopic Surgery has realized to the utmost those expectations which, but four years ago, were considered as altogether too sanguine. With pride can we say that this result is due to the diligence of Germans; and from our school at Vienna, in particular, an impetus has gone forth in all directions. Laryngoscopy has richly fulfilled what it promised. Its most brilliant achievement is, however, the extirpation of neoplasms. Every contribution in this direction is of worth and importance.

CASE 1.—A young man of about twenty-eight years of age, a lawyer by profession, came under my care in the spring. He had been wont to sing with great pleasure, and prided himself upon a beautiful tenor voice. For some time his voice, especially in the higher register, was slightly rough and shrill; the falsetto tones had become particularly bad. After any special exertion in singing, or in the practice of his profession, and after continued speaking without special effort, he would notice at evening an unpleasant feeling of heat, tension, and dryness in the pharynx, as well as a slight tickling and inclination to cough.

.FIG. 4.

An examination of both the pharynx and the larynx showed the existence of a sub-acute catarrh of these organs. The pharynx, especially, was injected whenever the patient became somewhat worse; the mucous membrane was glossy and dry, covered with a net-work of enlarged veins, and the uvula was slightly œdematous. The entrance to the larynx, as well as the ventricular and vocal chords, was reddened in fine streaks, some of which were at this time disappearing; the secretion of the mucous membrane was somewhat increased, and occasional small masses of a thin, yellowish mucus clung to the vocal chords. I commenced appropriate local treatment, blowing into the larynx a powder of alum and nitrate of silver, painting the pharynx· with iodine combined with glycerine, and also with a solution of nitrate of silver. I at first examined with the laryngoscope daily, and then every other day—partly for the purpose of applying remedies, and partly to follow the progress of the treatment. In the course of five or six weeks the laryngeal catarrh had vanished; that of the pharynx had so much improved, that the patient considered himself well, and, contrary to my advice, renewed his singing exercises. Our satisfaction did not last long.

After a fortnight had elapsed, my patient came again to me, and said there was still something wrong about his throat. I again examined him, and to my great astonishment found at

the first glance a small polyp about the size of a swollen millet-
seed, seated on the border of the anterior quarter of the left
vocal chord (Fig. 4); its surface was covered with mucus, was
pale-red and smooth; it was not movable, perhaps indeed from
its minuteness, nor did it change position during energetic inspiration and expiration, nor during the emission of sound, when,
however, it became pinched in the glottis.

There were in this case two circumstances which excited my
astonishment:

1st. That the patient, in spite of this formation, could sing
well; and of this I was convinced on hearing him, a few days
after, render the well known church aria from Alessandro Stra-
della. The selection was, to be sure, very favorable for his
voice; and no one, judging from the voice alone, would have
supposed that the larynx was diseased, or that a neoplasm
was seated upon the vocal chord. The voice seemed a little
uncertain, but that could have been explained on the ground of
a slight degree of embarrassment and of long interruption in
practice. The falsetto voice alone was altogether bad, a circumstance which seemed peculiarly adapted to throw some light on
the formation of the falsetto-register. As is well known, in spite
of many deviations, the writers upon the formation of the voice
have held with persistency that, in the chest voice, the vocal
chords vibrate in their entire extent, i.e. on their free edge, both
on the upper surface, the so-called ventricular, and on the lower
or sub-glottidean surface; on the other hand, the head voice
(falsetto) is produced by vibrations of the ventricular surface
and of the free edge alone. It is manifest, that, in our patient,
the formation of the chest voice was not much interfered with,
as, of three component factors, only one was wanting, while the
disturbance of the falsetto voice is comprehended when we
remember that, of its two factors, one—and according to many
writers the more important, viz. the vibration of the free edge
of the vocal chord—was wanting. Thus, it seems to me, this
case affords important confirmation of the views above mentioned
upon the physiology of the voice. Such observations are the
more important, as they can be made but rarely. Satisfactory
studies upon the formation of the voice, and especially upon the
differences of register, can only be made upon practised singers;
and it is a very rare coincidence that precisely such a one should

12

have a polyp upon the vocal chord. I leave physiologists to determine the value of my conclusions in this case.

2d. I was greatly astonished that I had so often, and, as I flatter myself, so carefully examined my patient, and that then, eight weeks after my first examination, and just after a fortnight's interruption, as soon as the mirror was introduced, I should for the first time discover the existence of the polyp, and with equal readiness at nearly every subsequent examination. The choice was left me to conclude that for six weeks I had examined carelessly, or that the neoplasm had been but very recently developed. The former supposition I cannot readily allow ; that I have the right to assume the latter, I cannot positively assert. The latter supposition is rendered probable from the subsequent growth of the formation. But either carries instruction with it.

The removal of a laryngeal polyp is indicated under all circumstances in this age of laryngoscopic operations, even when the polyp causes as little inconvenience as it did in the present case. I determined, therefore, to operate, and at once commenced the necessary preparations, viz. repeated introductions of the laryngeal sound in order to overcome the sensibility of the parts. I soon accomplished so much that the patient suffered no inconvenience when I lifted up the epiglottis, and, gliding down upon it, touched the polyp. This was the easiest method to bring the neoplasm into view. I postponed performing the operation, as my patient wished to make a month's tour among the mountains during September, to which I readily acceded in consideration of the apparent quiescence of the formation.

In the beginning of October we re-commenced treatment, and I gained by this delay the advantage of exhibiting the patient to my honored friend Prof. Wintrich, and of becoming acquainted with his modified *globe illumination*, which afforded me valuable assistance. In the meantime, at Wintrich's suggestion, Leiter had modified somewhat the blades of the laryngeal forceps, and produced a very useful instrument.

The operation was performed Oct. 25. [The author states that the wife of the patient was present, and aided greatly by her encouraging words in keeping the patient calm.—TR.] Although the patient could bear perfectly well the application of the laryngeal sound, and of Bruns's epiglottis forceps, still he urgently desired the application of local anæsthesia, after my worthy colleague Türck's method, and I did not oppose his wishes. I painted with a solution of morphia and chloroform the interior of the larynx by means of a mirror and a curved camel's-hair pencil, and the pharynx with the ordinary pencil used by artists, repeating the procedure at intervals at first of a minute, and afterwards as rapidly as possible. Although this operation was exceedingly unpleasant to the patient, it was still continued until I thought I detected a diminution of sensibility, which, indeed, did not amount to complete anæsthesia. This process of painting, however, developed one new fact opposed to the general belief. An almost insupportable degree of sensitiveness might have been anticipated from such long continued mal-treatment of the epiglottis with instruments, and with such irri-tating fluids as chloroform and rectified spirit, but such was not the case.

When I proceeded to the operation the mucous membrane of the larynx was somewhat reddened, and the larynx itself, when the epiglottis was well elevated, presented the appearance of Fig. 4 ; the neoplasm had increased since I first discovered it to the size of a hemp-seed. I had determined to seize the polyp with Voltolini's guillotine.*

After frequent futile attempts I succeeded' in catching the polyp in the ring, and in cutting off its free protruding half. Now the operation became so much the more difficult. Nume-rous attempts to catch the remnant with the guillotine signally failed. The forceps also were several times applied with no better success, and were finally thrown aside, in consequence of the weariness of all concerned, with the hope that the wounded polyp would perhaps take on retrograde action. The excised piece had been withdrawn sticking to the instrument, but it was lost ; the fork had not seized it.

This operation showed me the great defect in Voltolini's instrument. It would be well to have several of these instru-ments differently constructed ; for the prong upon the ring

* This instrument, like Bruns's scissors, is modelled after Charrière's pincette, with crossed arms. Each of the curved arms carries a small cutting ring of five decimètres in diameter. On pressing together the pincette, the rings shove over each other in such a way that a body presenting in the ring, is cut off. One of the rings carries on its lower surface a small prong with an upward direction, which pierces the body to be excised, and prevents its falling down.

must stand towards the free side, otherwise there is danger of perforating the vocal chord, and of doing it far more injury than the extirpation of the polyp could do good. It was a painful experience for me twice during the operation to meet with difficulty in withdrawing the instrument. Once, as reflex action was produced by touching the vocal chord, the glottis being closed spasmodically, the prong pierced the chord, and I could only free the instrument during the next deep inspiration, for which I had to wait. Again, in drawing out the instrument, the prong seized upon the left arytenoid cartilage. Further, a single instrument of this kind is only applicable with facility to one vocal chord; to the other it applies itself more or less obliquely, so that a protruding tumor cannot be seized exactly at its base, as in my own case: or else the operator must introduce the instrument with his left hand for the left vocal chord, a procedure which demands still greater facility of execution. The above remarks will be clearly understood, if we consider the position of the glottis in an individual sitting opposite to us. If therefore the guillotine is to prove a successful instrument, it must be so arranged that the ring shall turn in all directions. If the line $a\,b$ represents the left, and the line $a\,c$ the right vocal chord of a patient, it is manifest that (if we have the instrument in our right hand, and the mirror in our left) we can only apply the ring equally well to the right and to the left vocal chord, in case the ring admits of the adjustment referred to. Such an arrangement of the ring, however, is incompatible with the principle of the pincette. The same is true of Bruns's scissors, which, when introduced with the right hand, may indeed be placed parallel to the line $a\,c$ (the right vocal chord), but they always must stand more or less obliquely to the left vocal chord $a\,b$, thus incurring the danger of cutting the chord itself (which must by all means be avoided); otherwise we must practise using them with the left hand.

* * * * * * * * *

The expectation that the remnant of the polyp would shrink away of itself was not realized. The stump, however, rounded off, but still, after four weeks, preserved unaltered the form and size represented in Fig. 5. I determined, therefore, to repeat the operation, and commenced on the 26th of November, 1863, the same preparatory course as at the first. The application of the local anæsthetic had the same partial effect. I made use of the laryngeal forceps, to be described hereafter, and, after numerous fruitless attempts, I succeeded in seizing the polyp, and in

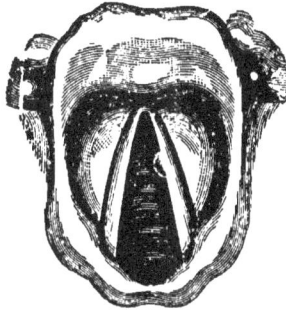

Fig. 5.

so far setting it free that it hung by only a few fibres. But this time there was to be no half-way work. When the patient uttered a continuous and half-suppressed æ, I could explore the closed glottis with my forceps, and after repeated efforts I thought I had actually seized the small nodule. That the prong of the forceps had pierced it, was not possible, for the body itself was too small, and I was most unpleasantly surprised not to find it clinging to the forceps or concealed in the hollow hemisphere. I seized the mirror and was about to make another attempt, and lo! there was a small lump about as large as the head of a middling-sized pin adhering to the surface of the mirror, enveloped in blood, and seeming quite compact when pressed between the fingers. It was the remnant of the neoplasm which had been thrown upon the mirror by the impulse of coughing, and had remained adhering to it. An examination showed at once that all the diseased part had been removed. The instrument had scraped away the epithelium from a large surface, especially from the left vocal chord. On both occasions the bleeding was not worth mentioning, and there was almost no reaction. At the first operation I had attempted, as I have stated above, to use the epiglottis forceps of Bruns; by these the epiglottis was several times severely pinched in spots as large as a lentil, and upon these places a yellowish exudation had been deposited.

Once the forceps had seized only the mucous membrane, and a circumscribed swelling was the result. By this injury of the epiglottis I easily explained the slight difficulty in swallowing and coughing, which lasted for a couple of days after the first operation, and, indeed, until the exudation upon the epiglottis had disappeared. The use of the morphia produced its general effects after the first operation; there was dizziness, sleepiness, headache, and nausea. No unpleasant effects, however, followed its use in the second case. A few days after, the epithelium had been entirely restored, and the slight injection and swelling upon the place of the operation had vanished. At present I cannot recognise, with any certainty, the spot where the polyp was seated.

The removed nodule was submitted to Dr. Schott (Rokitansky's first assistant—TR.) for microscopic examination, and was found to consist of areolar tissue with large loops of vessels and of epithelium.

The patient's head was slightly held by a trustworthy assistant in both operations. . . . . . For illumination I used a petroleum lamp with Wintrich's globe, and an operating spectacle with a concave mirror of nine centimètres diameter, and of eighteen centimètres focus. . . . . .

I learn with much pleasure that my patient has at last gathered courage to sing again, and that his voice has been found, by professional judges, to be as good as before the existence of the disease. The catarrh of the pharynx no longer exists. I do not desire to make the latter fact altogether dependent upon the operation, but some experience inclines me still to give due regard to this circumstance. The falsetto voice of my patient before the malady extended from F of the upper line up to B above, inclusive; these tones were lost during the existence of the polyp. Above B he could, both before and during the existence of the disease, make sounds, but they had no musical value. At present, five weeks after the operation, *the falsetto tones which were lost during the sickness are completely restored.*

The instruments which I have hitherto applied to operations upon polypi in the larynx are three—the forceps, a guillotine, with a sickle-shaped knife, and, finally, an instrument so arranged that any one of several blades may be introduced, the blade being at the same time concealed. As these instruments have

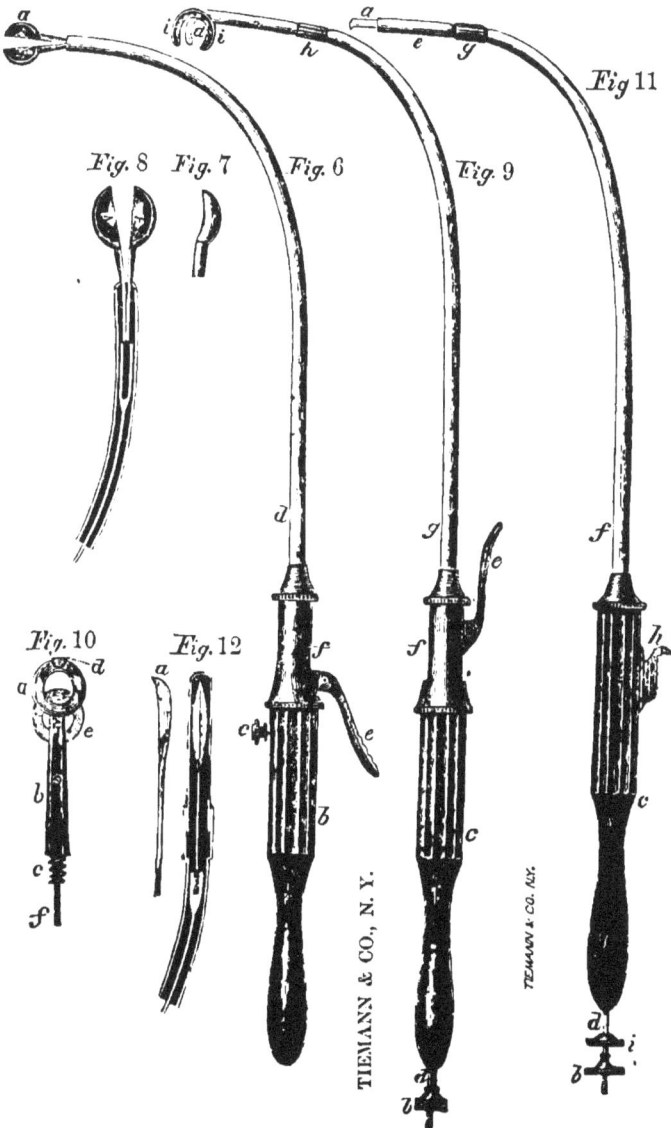

Fig 11

Fig. 8	Fig. 7	Fig. 6	Fig. 9

Fig. 10	Fig. 12

TIEMANN & CO., N. Y.

TIEMANN & CO. N.Y.

rendered good service, I will devote a few words to their explanation. Joseph Leiter, Instrument Maker and Bandagist, No
76 Alscr street, Vienna, can furnish any of them.

I. Leiter's Laryngeal Forceps

(Fig. 6, two-thirds of the actual size), is the result of manifold alterations and improvements.

The spring forceps, *a*, when closed, resembles the half of a hollow sphere; the
cutting edges, which shut upon one another (Fig. 7), are dull. Each blade of the
forceps has in its concavity a spear-point which does not extend beyond the cutting
edge. At the end of the spring the forceps terminates in a screw-thread (Fig. 8,
actual size), of six millimètres in length; by means of this the blades are firmly
screwed into a little tube at one end of a wire, the other end of which is firmly
held in the handle, *b* (Fig. 6), by the screw, *c*. The metallic tube, *d*, which can be
bent if necessary, moves over this wire and is shoved forward by pressure upon the
lever, *e*, and thus the blades of the forceps close quickly. When the lever, *e*, is
freed from pressure, the tube, *d*, is driven quickly back by a spiral spring placed
within the box, *f*, and thus the forceps are readily opened. The cutting edge of the
forceps, *a*, can be turned so as to seize a polyp in any direction. If the instrument
is to be taken apart the blades, *a*, must first be unscrewed, then the screw, *c*,
loosened, and thus the wire will be set free, and finally the milled head of the box,
f, must be unscrewed.

The rounded sickle-shaped knife invented by me (Fig. 9, two-
thirds the actual size), is intended to take the place of Bruns's
scissors. It may be applied in any direction.

The blade, *a* (Fig. 9), like the forceps in Fig. 8, is fastened to a wire, the end of
which, passing through the handle, *c*, is fastened by the screw, *b*, to a tube, *d*, which
moves in the handle. By a spiral spring concealed in the handle, *c*, the tube *d* is
pressed against the short fork-like end of the lever *e* which lies in the box, *f*. The
blade *a* is covered by a double two-leaved sheath, *i, i.* Both leaves of the sheath
are attached to a round tube, which terminates in a screw-thread fitting into the
double screw, *h.* This double screw *h* is fastened to, and turns upon the end of the
tube, *g*, which is stationary but which may be bent. If pressure is applied to the
lever *e* the knife slides suddenly out, and is again covered by the leaves of the
sheath. When the double screw *h* is unscrewed, the blade may be turned in any
direction, just as in the case of the forceps, *a*, in Fig. 6; then the blade being held
by its sheath in the left hand, the double screw is again tightened with the right,
and the instrument is ready for application.

Fig. 10 is a double ring between the leaves of which a ring-shaped blade plays.
This contrivance, resembling the Tonsillotome, may be screwed on to the instrument,
Fig. 9, instead of the sickle-knife, and it may be manipulated in the same manner.

Fig. 10, *a* represents the double ring, *b* its stem with its screw *c*; the two rings
are held in place on the one side by the clamp, *d*, and on the other by the screw at
b; *e* is the knife drawn down, and *f* its terminal screw.

The instruments, Figs. 9 and 10, may be taken apart by unscrewing first the
cutter, then the lever, finally the screw, *b*.

Neither the guillotine nor the sickle-shaped knife is provided with the spear for piercing a polyp. I have stated above the injurious effect which these spears may produce, and I think we attach altogether too much importance to the falling down into the air-passages of a small polyp or a fragment of one. The experience of Moura-Bourouillou confirms my opinion. He removed a small polyp from the vocal chord by means of a wire sling; on the end of the instrument there was a small arrow-head for piercing the tumor. But he had reckoned without the host, and the tumor fell into the trachea; it did not, however, cause the slightest inconvenience. In the case of larger formations this point would demand greater consideration, but in all such cases neither of these two instruments would come in question.

Fig. 11 (two-thirds the actual size) represents Leiter's covered knife, which may be applied in all directions, and into which cutting instruments may be introduced, either probe-pointed or lancet-shaped, or like an ordinary scalpel.

The attachment of the blade to the wire is accomplished as shown in Fig. 12 by a screw. The wire to which the blade is attached moves in the tube f and also in the handle c, in which it is fastened to the tube d by the screw b. The tube f is attached to the handle and does not admit of motion, but it may be bent. At its other extremity it has a contrivance for changing the direction (Fig. 12) like the one in the preceding instrument; only here the tube e, which is slit in front, serves as a sheath for the blade a. In Fig. 11 the double screw g holds the sheath fast. Fig. 12, a represents a scalpel-blade which may be introduced. The tube d is kept down by a spiral spring placed in the handle c, and thus the blade is kept concealed in the sheath. By shoving forward the slide h, which is screwed on to the tube d, the blade is driven forwards. The screw i serves to regulate the protrusion of the blade. The instrument may be taken to pieces by unscrewing first the knife, then the screws b and i, as well as the slide h, and finally the tube f.

The tubes of all these instruments are colored black, so as not to interfere with the reflex image in the mirror.

I can recommend these instruments from my own experience, and I believe that with these the operator would be quite well provided for all cases.

CASE II.—The case here recorded may indeed be considered as one of the most difficult for laryngoscopic operation.

A lady (her age I did not ask), a governess, came to me last autumn to be examined and to ask my advice. She had suffered for five years from complete aphonia, which had been gradually developed. On a careful examination, I discovered three formations of various sizes, as in Fig. 13. The largest of these was

spherical and was seated in the vicinity of the left vocal chord;
a second, smaller and club-shaped, projected from the anterior
angle of the glottis, and lay with its free extremity upon the
first; the third and smallest protruded from the anterior surface
of the right arytenoid cartilage, at about the level of the vocal
chord, and extended into this latter structure.*

Fig. 18.

I stated to her that I was inclined to operate; that I could
not insure success, so far as the restoration of the voice was concerned; but that by the operation, even as regarded the voice,
nothing was to be lost, since she was already voiceless, a circumstance sufficiently unfortunate for a governess. To my amazement, I must confess, there was no dyspnœa, not even an
abnormal murmur to be heard on auscultation.

[The author goes on to state that the patient had been examined by two laryngoscopists, one of whom made the diagnosis,
but at a time when the formation was not so large, and considered an operation impracticable; the other, however, looked
forward to an operation at a later date. The lady had also been
advised by another physician to go at once to London or Paris,
where he thought she might long ago have been freed from her
polyp; while in point of fact, up to that time, no laryngoscopic
operation had been undertaken in either of those cities.   Tr.]

The formation had at first, from its pale, reddish-yellow color,
from its dim lustre, and from its uneven raspberry-like surface,
led me to regard it as an enchondromatous or a fibroid tumor,
but as I studied it more carefully with reference to a future operation, I found by the aid of the sound that it had something of

* For convenience of reference these will be designated respectively as Nos. 1,
2, and 3. (Tr.)

the consistency of flesh, and, finally, I concluded that it must be composed of areolar tissue. The largest of the tumors proceeded, as I have said, from the left side of the ventricle of the larynx, and covered the left vocal chord in such a manner that only a small portion of the posterior part of the chord could be seen ; this fragment seemed to have a normal appearance. The anterior extremity of the left vocal chord was covered by the polyp No. 2. As the patient was so sensitive under the examination, I could not determine whether the polyp No. 1 lay free upon the left vocal chord, or whether it was intimately connected with it. When the glottis was closed all three of the polypi were shoved over each other. Thus much, however, I could observe. By the attempt to utter sounds, the polyp No. 1 was rolled up around its broad basis, so that it would then lie upon the right vocal chord, and would be wedged in between the ventricular and the vocal chords of both sides. The polypi Nos. 2 and 3 proved to be quite freely movable, following the respiratory current ; No. 2 particularly, with its free end glided down over No. 1, and was again thrown upwards by a forcible expiration.

The patient, aside from this local trouble, was apparently in perfect health, and there were no reasonable grounds for supposing a connexion between the local disease and any cachexia, or the preëxistence of any special disturbing cause.

On the first of November, 1863, I undertook the operation, after having forewarned the lady that she must have a large stock of patience, and must expect to undergo a second operation. The operation itself was undertaken after the same preliminaries as I have described in connexion with the first case, viz. the local application of morphia and chloroform, the fixation of the patient's head by a trustworthy assistant, and of the tongue by the patient's own fingers, Wintrich's globe apparatus on a petroleum lamp, the operating spectacles above mentioned, and the laryngeal mirror held loosely in the left hand. The efforts to produce anæsthesia were made at very short intervals for the space of two hours, but with very imperfect success; the epiglottis forceps could not be tolerated.

I removed, by means of the polyp forceps (see Fig. 6), properly adjusted, the formations Nos. 2 and 3, i. e. the one seated at the anterior angle of the glottis, and the one upon the right arytenoid cartilage, leaving of the former only a very small stump; accomplishing it, to be sure, only after frequent attempts. The greater part of the polyp No. 2 was removed by a single fortunate seizure; the wedge-shaped growth was about 1½ centimètres in length, and was first separated from its attachment, as the forceps which had seized it had reached the edge of the epiglottis.

Only a small portion of the polyp No. 3 was drawn out; the rest fell down into the trachea and was lost. The bleeding

FIG. 14.

FIG. 15.

and the pain were very slight; and there was scarcely any reaction. When the patient visited me on the third day after the operation, the mirror presented to me the appearance of Fig. 14. . . . With the closed glottis, as in the attempt to utter sounds, the appearance of Fig. 15 was presented. The polyp seemed to be quite hard. The patient was very much amazed that there was still no improvement in her voice, but I, for my part, could not be much surprised.

The microscopic examination of the portions removed, made by Dr. Schott, showed again newly developed areolar tissue, with large loops of vessels and epithelium.

On the 15th of November, 1863, I applied myself to the removal of the largest polyp No. 1, with the same preparations as before. I now knew that its consistency was quite compact, and was therefore convinced that the operation this time would prove very difficult; for it seemed probable beforehand that I should not be able to seize the polyp with the forceps on account of its remarkable size and rounded form. I determined, however, to make the attempt, and, if my suspicions proved correct, to divide the tumor with a knife, and finally remove it in fragments with the forceps. This soon became necessary. While the tumor, by a continued effort on the part of the patient to utter sound, was held firmly in the position shown in Fig. 15, I

succeeded, after one or two attempts, in making, with the aid of
the mirror, the two cuts represented in Fig. 14, using Leiter's
knife (Fig. 11), after having inserted a lancet blade.   The edges
of the cut bled but little, and they did not gape at all.

I then applied the forceps, and removed, piecemeal, the
greater part of this polyp.   On the larger of the removed
pieces, the smooth surface of the cut could be distinctly seen.
The bleeding was slight; altogether a couple of teaspoonfuls
might have been raised, little by little, mingled with mucus.
The operation proved very wearisome for all of us, having
lasted, inclusive of the attempts to produce anæsthesia, almost
four hours.   Still the reaction was limited to a slight pain in the
larynx, which lasted about three days.   The microscopic examination showed the same results as before.

When I next examined the patient, I found the appearance
represented in Fig. 16.   Of the large polyp there was only a

FIG. 16.

remnant of about the breadth of the vocal chord, with a projecting lobe on the posterior portion.   On the anterior angle of
the glottis, a small nodule had been reproduced.   To the astonishment of both myself and the patient, the voice unfortunately was not in the slightest degree improved.   I now began
to fear that the stump of the large polyp might have been so
attached' to the left vocal chord, or grown out of it in such
a manner, that the removal of the remnant without injury to
the vocal chord would be quite impossible.   So I contented
myself with blowing into the larynx most assiduously pulverized
alum, hoping to produce shrinking of the remnant of the polyp.
But it was all in vain; the voice was not restored.   In laugh-
ing as well as in quick inspiration, a short dull tone could be
elicited, but that was all.   Occasionally in speaking the sound

was somewhat rougher, but the change was only very transient. I now sought to destroy the stump of the polyp by cauterization, and twice applied nitrate of silver in substance; eschars were formed on the spots touched, but the voice did not come back. For me it was a painful moment; the patient must be encouraged to hope on, and I, for one, had nearly given up all hope. In the meantime, on the 31st of December, 1863, and on the 2d of January, 1864, I removed from the anterior angle of the glottis the lobe on the posterior extremity of the stump of the large polyp, and also the small growth which had recently sprouted. Both of these operations were conducted without assistance, and without any preliminary preparation. The lobe was as large as a small pea. I could not say why the voice did not yet return.

Meanwhile it occurred to me that when the patient had been left to herself for a couple of days, the interior of the larynx, especially the right vocal chord, was found somewhat pallid, and also that the sound of the voice was somewhat better. Hence I supposed that the larynx, in consequence of the perpetual blowing in of powder, the cauterizations, and the instrumental applications, must have been kept in a continual state of irritation, which of itself would at last affect injuriously the formation of the voice. I determined, therefore, to try non-interference with the larynx for a period of eight days, with the firm resolution at the end of that time, if no improvement had taken place, to resort to a combined operation with the knife and forceps. I was evident that this operation would now be much more diffi cult, as the stump was very narrow and flat, and hence that the line of my incision must be kept close to that of the ventricula: chord, while at the same time I must avoid injuring it; and ye if I should succeed in all this, and still find the polyp firml; attached to the vocal chord, all my labor would be in vain, an . the voice would be lost for ever. Although as an operator I ha, every ground to be satisfied with my experience thus far, yet, 1. the opinion of my patient and of the public, both I and my ai would be objects of reproach if the voice was not restored. Th: last trump was, therefore, to be carefully played.

I therefore recommended my patient, who coincided with a. my suggestions, to visit me again after the lapse of a few days, when we would come to a final conclusion. Now, I might as well confess that I was most agreeably surprised at the next visit, after an interval of several days, to hear my patient address me in a voice somewhat hollow and not metallic, but yet quite good and clear. The lady told me that for a year and a half she had not spoken so well as within the last two or three days, but that she was easily tired, and then her voice again became hoarse. An examination showed me that in consequence of the numerous and repeated applications, a superficial ulcer had been developed on the surface of the stump, from which I hope for a still further diminution of the remnant of the polyp. If the voice should not improve, and the patient should not be satisfied with its present condition, I shall be inclined to operate once more.

I think that this second case is as remarkable in its surgical as the first case was in its physiological relations. The "patience and perseverance" of the Germans, so wondered at by the French, and the German "cold-bloodedness" which astonishes them so much, have again won a triumph.

Czermak, to whom we owe all our progress in this department of surgery, can look back upon the last six years with contentment and pride. His name is for ever united with the history of laryngoscopy, and I am proud to have been his earliest student. To him I dedicate these lines with grateful affection and friendly regard.

DESCRIPTION OF THE PLATES.

PLATE I.

Fig. 1.—The Normal Naso-pharyngeal Space, the Catheter of Itard introduced into the Right Sinus of Rosenmüller, and into the Mouth of the
left Eustachian tube.

Fig. 2.—Catarrh of the Eustachian Tube.

Fig. 3.—Abnormal Growth upon the Septum.

Fig. 4.—A most Complete View of the Posterior Nares—the Vomer is
wanting.

Figs. 5 and 6.—Conditions observed after the Removal of Naso-pharyngeal
Polypi.

PLATE II.

Fig. 1.—Mucous Polypi on the Turbinated Bones of the left side.

Fig. 2.—Polyp of the Pharynx.

Fig. 3.—Ozœna Scrophulosa.

Figs. 4, 5, and 6.—Syphilitic Ulcerations in the Naso-pharyngeal Space.

Plate I

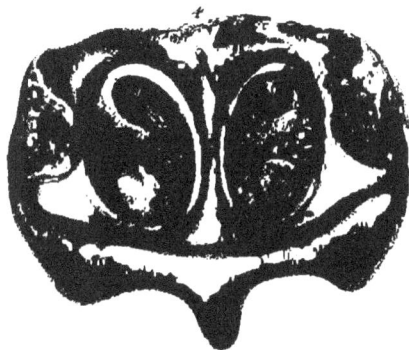

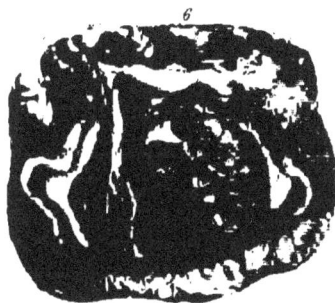

Plate II.

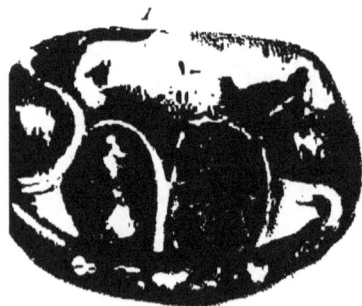
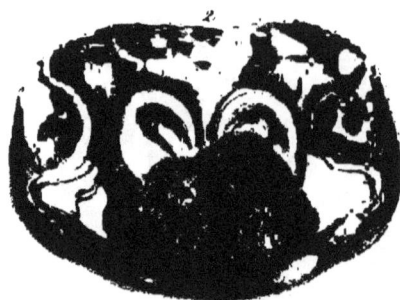

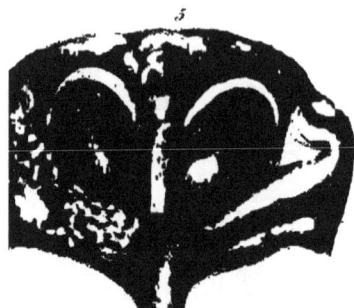

www.ingramcontent.com/pod-product-compliance
Lightning Source LLC
Chambersburg PA
CBHW021707210326
41599CB00013B/1557